ACPL ITEM
DISCARDED

DESIGN OF MODERN TRANSISTOR CIRCUITS

PRENTICE-HALL SERIES IN ELECTRONIC TECHNOLOGY

Dr. Irving L. Kosow, *editor*
Charles M. Thomson, Joseph J. Gershon,
and Joseph A. Labok, *consulting editors*

DESIGN OF MODERN TRANSISTOR CIRCUITS

MAURICE YUNIK

Department of Electrical Engineering
University of Manitoba
Winnipeg, Canada

PRENTICE-HALL, INC., *Englewood Cliffs, New Jersey*

Library of Congress Cataloging in Publication Data

YUNIK, MAURICE.
  Design of modern transistor circuits.

  (Prentice-Hall series in electronic technology)
  1. Transistor circuits. 2. Electronic circuit design. I. Title.
TK7871.9Y85   621.3815′3′0422   72-6047
ISBN 0-13-201285-5

© 1973 by Prentice-Hall, Inc.
Englewood Cliffs, New Jersey

All rights reserved. No part of this book may be reproduced in any form or by any means without permission in writing from the publisher.

Printed in the United States of America

10 9 8 7 6 5 4 3 2 1

PRENTICE-HALL INTERNATIONAL, INC., *London*
PRENTICE-HALL OF AUSTRALIA, PTY., LTD., *Sydney*
PRENTICE-HALL OF CANADA, LTD., *Toronto*
PRENTICE-HALL OF INDIA (PRIVATE) LTD., *New Delhi*
PRENTICE-HALL OF JAPAN, INC., *Tokyo*

1742234

*To my wife
and my parents*

CONTENTS

Preface   xiii

CHAPTER 1

Essential Points of Circuit Theory   1

    1.1   Ideal Sources   1
    1.2   Network Theorems   3
    1.3   Voltage and Current Division in Resistive Networks   7
    1.4   Practical Sources   7
    1.5   Superposition   10
    1.6   Input and Output Impedance   12

CHAPTER 2

Transistor as Amplifier   16

    2.1   Terminal Behavior of Transistors   16
    2.2   Input Behavior of a Transistor   19
    2.3   Symbols for a Transistor   20
    2.4   General Amplifiers   21
    2.5   Transistor as Current Amplifier   22
    2.6   Symbols   24
    2.7   Transistor as Voltage Amplifier   25

CHAPTER 3

Two-Port Parameters
and Two-Port Equivalent Circuits 33

    3.1   Definitions and Terminology for Two-Ports   33
    3.2   Types of Two-Port Parameters   34
    3.3   Hybrid Equivalent Circuit   35
    3.4   Characteristic Curves and h-Parameters   40
    3.5   Dynamic Measurement of h-Parameters   43
    3.6   z-Parameters and the Equivalent Circuit   47

CHAPTER 4

Biased Common Emitter Amplifier 52

    4.1   Constant Current Biasing   52
    4.2   H-type Biasing   55
    4.3   Current-Feedback Biasing   58
    4.4   Use of Equivalent Circuit
           for a Transistor in Biased State   60
    4.5   Voltage Gain, Input and Output Impedance
           of a Current Biased Amplifier   61
    4.6   Load Line Analysis   68
    4.7   H-type Biased Amplifier   73
    4.8   Unbypassed Emitter Resistor Amplifier   75

CHAPTER 5

Generic Equivalent Circuit 85

    5.1   Semiconductor Diode   85
    5.2   Use of Semiconductor Diode   92
    5.3   Extension of Diode Equation
           to the Transistor   97
    5.4   Generic and Tee Models for a Transistor   100
    5.5   h- and r-Parameter Interrelationships   104
    5.6   Properties of Common Emitter Amplifier
           Using Tee Model   109

CHAPTER 6

Common Base and Common Collector Amplifiers — 115

- 6.1 Biasing the CB Amplifier   115
- 6.2 Electronic Properties of the CB Amplifier   116
- 6.3 Two-Stage Amplifiers   122
- 6.4 Common Collector (Emitter Follower) Amplifier   128
- 6.5 Load Line Analysis for the Emitter Follower   135
- 6.6 Bootstrapped Amplifiers   136
- 6.7 Design of Amplifier Circuits   142
- 6.8 Some Examples of Unusual Cascading   142

CHAPTER 7

Field Effect Transistor — 155

- 7.1 Operation of the Junction FET (JFET)   155
- 7.2 Insulated Gate FET (IGFET)   159
- 7.3 Equivalent Circuit for the FET   161
- 7.4 Biasing the FET   164
- 7.5 Small Signal Analysis of an FET Amplifier   168
- 7.6 Common Drain Amplifier (Source Follower)   171
- 7.7 Common Gate Amplifier   175
- 7.8 FET as a Voltage Variable Resistor   176

CHAPTER 8

Aids for Design — 181

- 8.1 Circuit Specifications   181
- 8.2 Transistor Specifications   182
- 8.3 Logarithmic Gain   188
- 8.4 Use of Frequency-Response Plots   189
- 8.5 Bode Frequency-Response Plots   192

CHAPTER 9

Low-Frequency Response of Transistor Amplifiers — 201

- 9.1 High-Pass R–C Filter   201
- 9.2 Input Coupling Capacitor   203
- 9.3 Emitter Bypass Capacitor   205
- 9.4 Multistage Capacitive Coupling   210
- 9.5 Combined Effects of the Two Coupling Capacitors   211
- 9.6 Combined Effects of the Coupling and the Bypass Capacitors   214
- 9.7 Frequency Effects in an FET due to Coupling and Bypass Capacitors   217
- 9.8 Transformers for Transistor Circuits   222
- 9.9 Analysis of Transformer Coupled Amplifiers   229
- 9.10 Design of Transformer Coupled Amplifiers   231

CHAPTER 10

High-Frequency Response of Transistor Amplifiers — 239

- 10.1 Hybrid-Pi Equivalent Circuit   239
- 10.2 β Cutoff, α Cutoff, and Gain-Bandwidth Product   247
- 10.3 Miller Effect   251
- 10.4 Total Frequency Response of the CE Amplifier   260
- 10.5 High-Frequency Response of the CB Amplifier   264
- 10.6 High-Frequency Response of the Cascade Amplifier   270
- 10.7 Frequency Response of the Emitter Follower   273
- 10.8 Frequency Response of the Field Effect Transistor   278

CHAPTER 11

Power Supplies — 285

- 11.1 Rectification   285
- 11.2 Ripple and Regulation   289
- 11.3 The R–C Filtered Rectifier   290
- 11.4 Emitter Follower Voltage Regulator   293
- 11.5 Zener Diode Regulator   296
- 11.6 Regulated Power Supply Design   302

CHAPTER 12

Feedback Amplifiers and Oscillators   **309**

    12.1 Basic Feedback Theory   309
    12.2 Circuits with Negative Voltage Feedback   315
    12.3 Additional Advantages of Negative Feedback   318
    12.4 Current-Feedback Amplifiers   321
    12.5 Stability of Feedback Amplifiers:
          Nyquist Criterion   326
    12.6 Sinusoidal Oscillators
          and the Barkhausen Stability Criterion   327
    12.7 Practical Oscillators   329

CHAPTER 13

Audio Power Amplifiers   **340**

    13.1 Power Considerations   341
    13.2 Classification of Power Amplifiers   345
    13.3 Class B Push–Pull Amplifiers   346
    13.4 Load Line Analysis of Class B Amplifiers   350
    13.5 Practical Push–Pull Amplifiers   352

Index   **359**

PREFACE

This book is intended to teach students to design actual electronic circuits in a practical, down to earth manner. It is designed as a first text in electronic circuits for both the university and technical school student. Its primary emphasis is on the design aspects of electronic circuits. Therefore, it should be valuable to practicing design engineers as well.

The text begins with a partial review of circuit theory which serves to emphasize the approximation aspects of circuit calculations. This material is not generally covered in most circuit theory texts. However, it is assumed that the student has a working knowledge of ac and dc circuits.

Design of simple circuits are presented as early as possible (Chapter 2) to provide motivation for the new student. Here a simple but accurate discription of biasing and amplifier operation is given. This is followed by a more detailed approach in chapters 3 through 6, using h and r parameters. Transistors are treated from a terminal behavior point of view with the exception of Chapter 5 where a descriptive treatment of the physics of solid state devices is presented. Chapter 5 begins with only those results from solid state physics which are of immediate interest to the circuit designer. The remainder of the chapter then employs these results to embellish the theory obtained from using the terminal point of view in the three previous chapters. The field effect transistor is introduced in Chapter 7. Chapters 8 through 13 deal with techniques of circuits such as; amplifiers, power supplies, regulators, oscillators, etc. The objective in these chapters is to develop, by means of circuit analysis, formulae which aid the student in designing practical circuits.

This book emphasizes the fundamentals of circuit theory and its application in the design of practical circuits. An attempt has been made to provide many additional comments for the student which would not normally be found in electronics books, in order to motivate the student toward the design aspect of electronics.

I am indebted to many colleagues for their help and, especially, to all the students who have helped with their questions, comments and criticisms throughout my teaching experience. I would like to mention especially my former teacher, Professor A. Simeon, and Professor J.P.C. McMath. The staff of Prentice-Hall has been most cooperative and have contributed greatly to the manuscript. Special thanks are extended to Mrs. Jean Heinamaki who typed most of the manuscript.

MAURICE YUNIK

Manitoba, Canada

DESIGN
OF
MODERN TRANSISTOR
CIRCUITS

1

ESSENTIAL POINTS OF CIRCUIT THEORY

To design electronic circuits successfully we must have, in addition to a knowledge of the basic tools of circuit theory, a knowledge of how and when to make reasonable approximations. Besides simplifying the designer's work, appropriate approximations usually lead to a better insight into the actual operation of circuits and provide the designer with an intuitive and often more direct approach to his design. Approximations enable him to adjust the more predominant variables.

In this process of making such approximations, however, the designer must never lose sight of the definitions and results of the very fundamental tools of circuit theory.

Thus our study begins with a review of some simple fundamental idealizations from circuit theory and the use of approximations to simplify the manipulative labor.

1.1 Ideal Sources

An *ideal voltage source* may be defined as a device that delivers a defined voltage across its terminals independently of what is connected to these terminals. The symbol for an ideal voltage source is shown in Fig. 1-1a.

An *ideal current source* is defined as a device that delivers a defined current from its terminals regardless of the nature of the circuit connected to it. The symbol for an ideal current source is shown in Fig. 1-1b.

Let us now observe some of the implications of these ideal devices. The

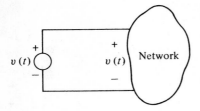

(a) Ideal voltage source

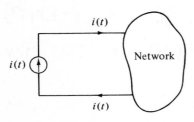

(b) Ideal current source

FIGURE 1-1 Symbols showing ideal voltage and current sources.

following ideas may seen quite obvious, but they have very important consequences in the understanding of transistor circuits. First of all, let us connect a single variable resistor to an ideal voltage source, as shown in Fig. 1-2a, and vary it in some fashion from a very small value to a large value. Notice

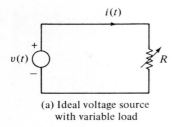

(a) Ideal voltage source with variable load

(b) Ideal current source with variable load

FIGURE 1-2 Action of ideal sources on variable resistor.

that the voltage (from our definition) remains the same and the current, $i(t)$, varies from a very large value to a very small value.

Next, let us connect this same resistor to a current source, as shown in Fig. 1-2b.* Here we see that the voltage becomes larger and larger as we increase the value of this resistor. It is this very property that makes a transistor operate in the fashion that it does, since the output of a transistor is almost like an ideal current source.

A current source is somewhat harder to visualize at first, since ideal voltage sources seem to be more common—such as a 110-volt (V) supply we use for electrical distribution. However, even this is not an ideal voltage source, since, if the load added to it is made large enough, the terminal voltage will drop. Thus, to bring us out of the idealized world, we should next consider some practical sources. Before doing so, however, let us first consider two very important theorems, which follow.

1.2 Network Theorems

Thévenin's theorem states the following: Any linear active network may be represented by an equivalent network consisting of a voltage source equal to the open-circuit voltage of the network in series with the original network, with all independent sources set to zero.

To set sources equal to zero we merely short-circuit all the *voltage* sources and open-circuit all the *current* sources. This process has an intuitive appeal, since a short circuit (a wire) has no voltage across it and hence a voltage source of zero volts. Similarly, an open circuit (a break) has a zero current flowing through it and may be thought of as current source of zero amperes. A pictorial representation of Thévenin's theorem is shown in Fig. 1-3. Here we obtain the Thévenin equivalent of network A, which is connected to network B by two leads. Notice that the active network A can be part of a larger network, as shown in Fig. 1-3a, and that v_θ, the open-circuit voltage, must be placed in the same direction as the one in which the measurement was done as shown in Fig. 1-3c.

To illustrate the use of this theorem let us consider a few examples, which will also incorporate reasonable approximations. In electronics it is reasonable to approximate within 5 per cent or sometimes as much as 10 per cent, since most components, as specified by manufacturers, have these types of tolerances. For example, the most common resistors are specified within 10 per cent of their given values.

---

*The sign convention adopted is a + sign for voltage at the tail of the current arrow for passive elements.

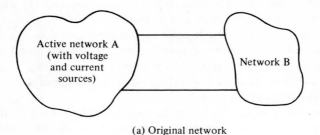

(a) Original network

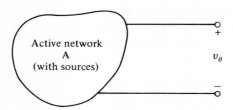

(b) Voltage equivalent portion

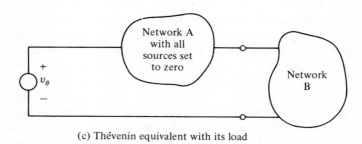

(c) Thévenin equivalent with its load

FIGURE 1-3 Decomposition of a network to its Thévenin equivalent.

Example 1-1: For the circuit shown in Fig. 1-4a, (a) find the Thévenin equivalent to the left of terminals $a - b$, and (b) proceed to find v_2.

Solution:

(a) The first step is to combine R_2 and R_3 to a single resistor and call it R_6. Then

$$R_6 = \frac{R_2 R_3}{R_2 + R_3} = \frac{5 \times 50}{5 + 50} \text{ k}\Omega \cong 5 \text{ k}\Omega$$

Notice that $R_6 = 5 \text{ k}\Omega$, since the effect of the 50-kΩ resistor in parallel with the 5-kΩ resistor is negligible.

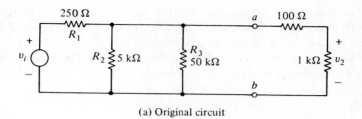

(a) Original circuit

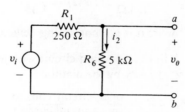

(b) Equivalent Thévenin voltage

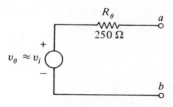

(c) Thévenin equivalent

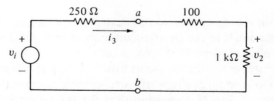

(d) Thévenin equivalent with load

FIGURE 1-4 Solution of Ex. 1-1 using Thévenin's theorem and a series of reasonable approximations.

To find v_θ we open-circuit terminals $a - b$, as shown in Fig. 1-4b. Now

$$i_2 = \frac{v_i}{5 \text{ k}\Omega + 250 \text{ }\Omega} \cong \frac{v_i}{5 \text{ k}\Omega}$$

and

$$v_\theta = i_2 \, 5 \text{ k}\Omega \cong v_i$$

In other words, the open-circuit voltage $v_\theta = v_i$, since the voltage drop across R_1 is negligible. These are reasonable approximations to make in electronics. Notice that a similar approximation is made to obtain R_θ in Fig. 1-4c. Placing R_θ in series with v_θ leads us finally to the Thévenin equivalent.

(b) To obtain v_2, we find i_3 and proceed:

$$i_3 = \frac{v_i}{1350 \; \Omega}$$

$$v_2 \simeq \frac{1}{1.35 \; k\Omega} v_i (1 \; k\Omega) = \frac{1}{1.35} v_i$$

It is apparent that this theorem is a labor-saving device.

If we call the network in the Thévenin circuit with all sources set to zero Z_θ, then we can find Z_θ very easily by the method

$$Z_\theta = \frac{v_\theta}{i_{sc}} = \frac{\text{open-circuit voltage}}{\text{short-circuit current}} \tag{1-1}$$

This procedure may be justified by considering Fig. 1-5. Every network can be represented as shown here according to Thévenin's theorem, and it is obvious that Z_θ is given by the above expression.

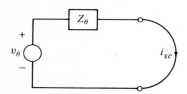

FIGURE 1-5 Method for determining the Norton equivalent from a Thévenin equivalent circuit.

Norton's theorem states that every linear network can be represented by a current source equal to the short-circuit current in parallel with the same network, with all the independent sources set to zero.

It is left as an exercise for the student to show that this is true (it follows from Thévenin's theorem) and to work out a suitable example to show the use of such a theorem. The Norton equivalent for the network in Fig. 1-5 is shown in Fig. 1-6. Hence we can find the Norton equivalent easily by first finding the Thévenin equivalent, although this may not always be the most convenient way.

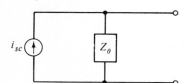

FIGURE 1-6 A Norton equivalent circuit.

1.3 Voltage and Current Division in Resistive Networks

Although the following equations do not warrant the name theorem, the author often refers to them in this manner. These "theorems" are actually labor-saving devices for quick calculations.

Consider the situation in Fig. 1-7, where a series connection of resistors is made. Then it follows that

$$i = \frac{v_i}{R_1 + R_2 + \ldots + R_n} \tag{1-2}$$

and

$$v_j = iR_j = \frac{R_j}{R_1 + R_2 + \ldots + R_n} v_i \tag{1-3}$$

Stating the results in words, we can say that the voltage drop across a resistor R in series with other resistors is equal to R divided by the total resistance in the circuit multiplied by the applied voltage.

Similarly, for a parallel setup, as shown in Fig. 1-8, it follows that i_j, the current in the j-branch, is given by

$$i_j = \frac{G_j}{G_1 + G_2 + \ldots + G_n} i \tag{1-4}$$

Where the Gs are the conductances in the circuit in mhos.

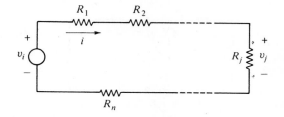

FIGURE 1-7 Network illustrating voltage division.

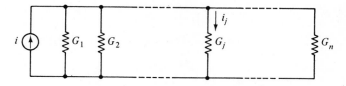

FIGURE 1-8 Network illustrating current division.

1.4 Practical Sources

Since few if any actual ideal voltage or current sources arise in practice, it is useful to define a *practical voltage* and a *practical current source*.

8 Chap. 1 / Essential Points of Circuit Theory

Consider the situation in Fig. 1-9a. The circuit in the dashed box represents the Thévenin equivalent circuit of a signal generator whose output impedance is R_G. As long as R_L, which is a load connected to the generator, is much larger than R_G, we can see by voltage division that the load sees approximately the voltage v_G even when R_L is varied. If this is true for all R_L that we will use, we shall call the signal generator a *practical voltage source* or an *approximate voltage source*.

A similar definition can be set up for a *practical* or *approximate current source* in the following manner. Consider the situation in Fig. 1-9b, where $R_G \gg R_L$, a signal generator with a high output impedance. Then

$$i(t) = \frac{v_i}{R_G + R_L} \simeq \frac{v_i}{R_G} \tag{1-5}$$

since $R_G > R_L$. Hence the current $i(t)$ through R_L remains invariant with various values of R_L as long as $R_L < R_G$.

An alternative way of looking at a practical current source is to consider the Norton equivalent of Fig. 1-9b, as shown in Fig. 1-10. Notice that $i_2 \simeq i(t)$, since by current division very little current flows through R_G as long as $R_G > R_L$. This alternative technique of looking at a current source is important in understanding the operating of transistors, since the output of a

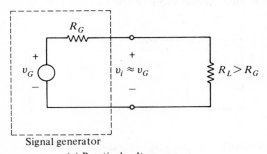

(a) Practical voltage source

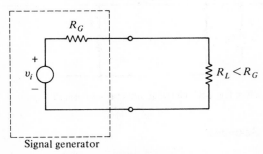

(b) Practical current source

FIGURE 1-9 Practical voltage and current sources.

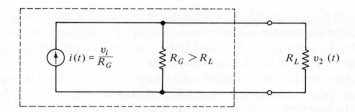

FIGURE 1-10 An alternate visualization of a practical current source.

transistor is the same as that shown in Fig. 1-10. It is apparent that increasing R_L increases the voltage $v_2(t)$.

A practical voltage source may also be created in a laboratory if the condition that $R_G < R_L$ is not met. A large resistor R_{G2} is connected in series with a small resistor R_{G3} to the output of the signal generator, as shown in Fig. 1-11. If $R_{G3} \ll R_{G2}$, then v_2 remain invariant with R_L, provided R_L is larger than R_{G3}. v_2 can be measured with a voltmeter and will be the Thévenin voltage in series with R_{G3}, since R_{G2} is large. But R_{G3} is small, and therefore we have created an approximately ideal voltage source. Figure 1-12 shows the Thévenin equivalent to the left of the terminals a-b (according to suitable approximations).

A practical current source may be created by placing a resistor R_{G4} in

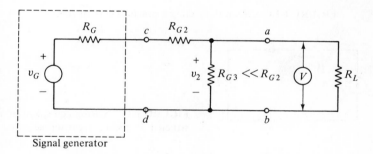

FIGURE 1-11 Construction of practical voltage source.

$$v_2 = \left(\frac{R'_{G3}}{R_G + R_{G2} + R_{G3}} \right) \times v_G$$

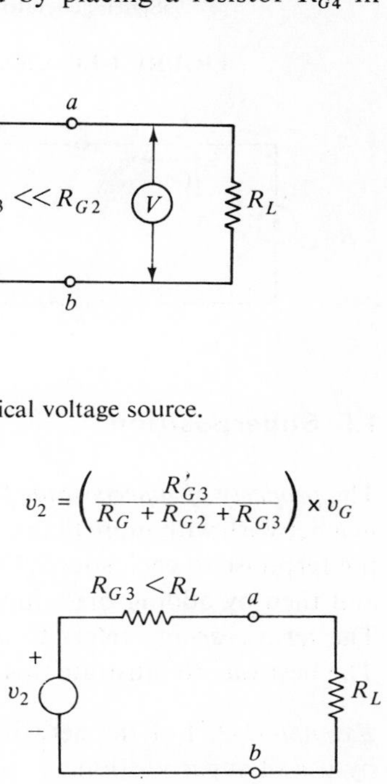

FIGURE 1-12 Thévenin equivalent of a constructed practical voltage source.

10 Chap. 1 / Essential Points of Circuit Theory

series with the signal generator whereby $R_{G4} > R_L$ and R_G. It follows then that the current created is approximately equal to v_G/R_{G4} and remains invariant with R_L, since $R_{G4} < R_L$. See Fig. 1-13, where you will notice that Figs. 1-13 and 1-14 are approximately equivalent.

Another way to create an ideal voltage source in a laboratory with a signal generator that has an output impedance Z_G and is connected to a load R_L is to make sure that the load R_L receives a constant voltage during the course of the experiment. This may mean that the amplitude of the signal generator may have to be changed as R_L changes. The justification for this procedure is that it satisfies the definition and so R_L sees an ideal voltage source.

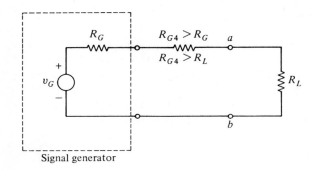

FIGURE 1-13 Circuit illustrating a practical current source.

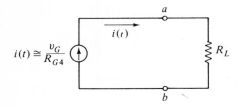

FIGURE 1-14 Norton equivalent of a constructed practical current source.

1.5 Superposition

The *property of superposition* for linear networks states that the total response in a network with more than one source may be determined first by computing the response to each source (by individually setting all other sources to zero) and then by adding these individual responses to obtain the total response. The term *response* refers to a voltage or current elsewhere in the network. The best way to illustrate this property is by a numerical example.

Example 1-2: For the network shown in Fig. 1-15, calculate the voltage v_2 by use of superposition.

Sec. 1.5 / Superposition   11

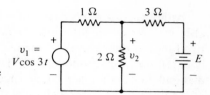

FIGURE 1-15 Example of a circuit where the property of superposition may be employed.

Solution:

First of all, set the battery E equal to zero by replacing it by a short circuit and find v'_2. The voltage v'_2 is the response due to v_1, acting alone in the circuit. Let us also convert v_1 and the 1-ohm(Ω) resistor to current source by use of Norton's theorem and solve for the current i'_2 in the 2-Ω resistor by current division, as shown in Fig. 1-16. Hence

$$i'_2 = \frac{\frac{1}{2}}{1 + \frac{1}{2} + \frac{1}{3}} V \cos 3t$$

$$= \frac{3}{6 + 3 + 2} V \cos 3t$$

$$= \tfrac{3}{11} V \cos 3t$$

and
$$v'_2 = 2i'_2 = \tfrac{6}{11} V \cos 3t$$

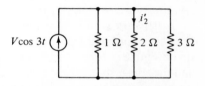

FIGURE 1-16 Partial solution of Ex. 1-2 using superposition.

In the next step we set v_1 equal to zero by replacing it by short circuit and solve for v''_2, the voltage due to E alone, in a similar fashion, as shown in Fig. 1-17. Then

$$i''_2 = \frac{\frac{1}{2}}{1 + \frac{1}{2} + \frac{1}{3}} \frac{E}{3}$$

$$= \frac{E}{6 + 3 + 2}$$

and
$$v''_2 = \tfrac{2}{11} E$$

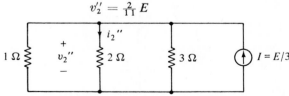

FIGURE 1-17 Remainder of solution for Ex. 1-2 using superposition.

Finally, by superposition, $v_2 = v'_2 + v''_2$, and

$$v_2 = \tfrac{2}{11} E + \tfrac{3}{11} V \cos 3t$$

Although this result could have been calculated more directly and probably with less difficulty by the application of Kirchhoff's voltage law to two loops in the network, Example 1-2 illustrates the type of thinking that is necessary in electronic design. This technique may save many laborious manipulations in a much larger circuit. Another reason for the importance of the property of superposition is that partial solutions due to time-varying signals may be obtained in a network. If v_1 is a signal voltage, then we may be interested in the amount of signal that is present in another part of the circuit and not at all interested in the voltage due to a battery. This property allows us to solve only for the time-varying or the signal voltage elsewhere in the circuit.

1.6 Input and Output Impedance

An electrical network with a single set of terminals is called a one-port network. This type of network can be driven by either a voltage or current source, and voltage and current are related by the impedance of the network. For purely resistive networks the voltage and current are related through Ohm's law, but for a general network with a sinusoidal source the voltage–current relationships are found by the use of phasors.

A network with two sets of terminals is called a *two-port*. The terminals may be named the *input* and the *output* terminals. The name input implies that a source (a time-varying signal perhaps) is applied at the input terminals and that the response or the desired signal is obtained at the output terminals. An example of a two-port is illustrated in Fig. 1-18.

For a purely resistive network the input voltage and input current are related by Ohm's law and can be calculated if the equivalent resistance of the network is known. Let us call this resistance R_{in} (the "in" subscript used to designate input), which is the input resistance or impedance of the network. For the circuit shown in Fig. 1-19

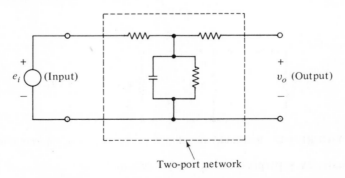

FIGURE 1-18 Example of a two-port network.

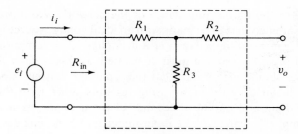

FIGURE 1-19 Example of a two-port network for which the input impedance is to be calculated.

$$R_{in} = R_1 + R_3$$

and
$$i_i = \frac{e_i}{R_1 + R_3} = \frac{e_i}{R_{in}}$$

In many circuits the internal components of resistive networks cannot be measured individually. Suppose that R_1, R_2, and R_3 in Fig. 1-19 are three resistors enclosed in a sealed box; then R_{in} cannot be calculated from a knowledge of R_1 and R_2 and must be measured.

To measure R_{in} we obtain a signal source e_s (a practical source), and place a variable resistor R_v in series with the two-port network and the source, as shown in Fig. 1-20. Notice that the practical source has a resistance R_s associated with it. An ideal voltage source can be easily created by keeping e_i a constant during the course of the measurement if R_s is not negligible with respect to R_v and R_{in}. R_v is now adjusted until e_{in} is equal to $\frac{1}{2} e_i$. Hence R_v is equal to R_{in}, since they have equal voltages across them.

Since the entire circuit in Fig. 1-19, including the source, can be represented by a Thévenin equivalent circuit, the impedance in series with Thévenin source can be thought of as an output impedance of a circuit. This impedance can be easily measured by removing R_v from the input and placing it across the output. If R_v is adjusted so that the voltage across it is one half the open-circuit voltage at the output, then R_v is equal to the Thévenin impedance or

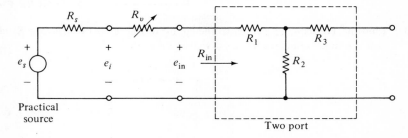

FIGURE 1-20 Technique used to measure the input impedance of a two-port network.

simply equal to the output impedance of the network. We must remember that the measurement has to be made with an ideal voltage source at the input. If the signal generator impedance is not negligible in comparison to R_{in}, then a practical voltage source must be created, as discussed in Section 1-4.

The reader should also notice that the input and output impedance can be obtained in an experiment by using resistors other than those exactly equal to the input and output impedance of the network, as shown in Fig. 1-20 for example. To calculate the input impedance here, one determines the current through R_v and uses e_{in} to obtain R_{in}.

PROBLEMS

1.1 (a) In Fig. P1-1 solve for $i(t)$ if $v(t) = 3$ V.
(b) What will $v_2(t)$ be if $v(t) = 6 \cos(200\pi t)$? Sketch the wave form of $v_2(t)$.

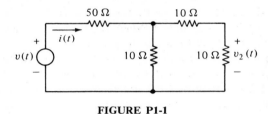

FIGURE P1-1

1.2 (a) Find the Thévenin equivalent in Fig. P1-2. Use reasonable approximations to obtain your result.
(b) What is the Norton equivalent of this circuit?

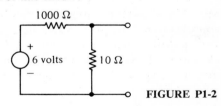

FIGURE P1-2

1.3 Find the Thévenin equivalent of the circuit in Fig. P1-3 by applying Kirchhoff's voltage and current laws directly to the circuit.

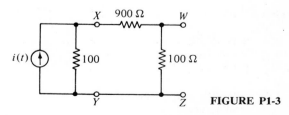

FIGURE P1-3

1.4 Find the Thévenin equivalent of the circuit in Fig. P1-3 by first finding the Thévenin equivalent of the circuit to the left of terminals $X - Y$.

1.5 Show that, if 1-, 5-, and 10-kΩ resistors are applied to terminals W-Z in Fig. P1-3, the circuit acts as if it were a practical voltage source.

1.6 (a) Obtain the Norton equivalent of Fig. P1-6 by using current division to obtain the short-circuit current. Use reasonable approximations to obtain your result.
(b) Show that this circuit acts as a practical current source to 1-, 2-, or 3-Ω resistors when connected at the terminals X-Y.
(c) Show also that the circuit acts as a practical voltage source if 2-, 5-, and 10-kΩ resistors are connected to terminals X-Y.

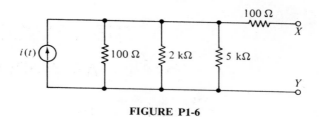

FIGURE P1-6

1.7 Obtain $i(t)$ in Fig. P1-7 by the use of Kirchhoff's voltage and current laws.

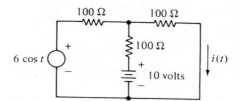

FIGURE P1-7

1.8 Repeat Problem 1.7 by using the property of superposition.

1.9 Find the input impedance of the network in Fig. P1-9 by applying a voltage $v(t)$ to terminals X-Y and finding the current into the network.

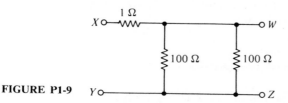

FIGURE P1-9

1.10 In Fig. P1-9, what is the output impedance of the network if a signal generator with an internal resistance of 100 Ω is connected to the terminals X-Y?

2

THE TRANSISTOR AS AMPLIFIER

The commonplace transistor is a three-terminal device that can be used to amplify voltages or currents. The transistor is made from semiconductor material, such as germanium or silicon. Since its birth in 1948, the transistor has become very popular because of its small size, extreme ruggedness, and low voltage operation, so that its forerunner, the vacuum tube, has almost become obsolete.

Since the two fundamental quantities that can be easily measured in the laboratory are voltage and current, we shall study the transistor from the point of view of its terminal behavior. The transistor is not a linear device and does not obey Ohm's law; that is, the voltages and currents at a particular set of terminals are not related by a constant. Thus a study of the relationships between voltages and currents at the terminals of a transistor is necessary to learn about its complete behavior.

2.1 Terminal Behavior of Transistors

Let us consider the setup shown in Fig. 2-1. The leads on the transistor are named the base, emitter, and collector and are labeled B, E, and C, respectively. Notice that although the transistor has three leads, it is operated as a two-port with the emitter terminal common to the input and output. This type of setup is called the common emitter (CE) amplifier. V_{BB} and V_{CC}, the batteries (or *dc* supplies), supply the input and output with power. V_{CC}, V_{BB}, and the load are adjustable to obtain various values of v_{BE}, i_B, v_{CE}, and i_C,

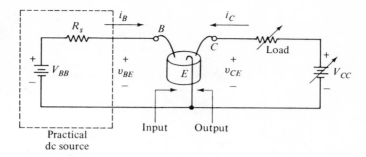

FIGURE 2-1 A common emitter amplifier.

the input and output voltages and currents. Since a transistor is nonlinear, its action is most easily interpreted graphically, as shown in Fig. 2-2a. These curves are called the *static output* or *collector characteristics*. The values given in the figure are typical of a small-signal transistor, but the magnitude of the voltages and currents varies considerably from transistor to transistor.

The transistor is basically a current-amplifying device. Notice that when i_B is equal to zero (open circuit at the input), little or no output current flows, and for most transistors, this current is only about 10 or 20 microamperes (μA). If one increases the input current to a value of 10μA by adjusting V_{BB} and using the appropriate value of R_s, considerable collector current i_C flows in the direction indicated in Fig. 2-1. Notice in Fig. 2-2a that the collector current is almost independent of the value of v_{CE}. This means that although V_{CC} and the load can be changed to many different values, the same i_C flows. For this transistor in Fig. 2-2a we get about 1 milliampere (mA) of collector current when 10 μA of base current is applied. An increase or decrease in i_B, however, gives an approximately proportional increase or decrease in i_C. The proportionality constant relating i_C and i_B is called h_{FE}, the *forward current amplification factor*:

$$h_{FE} = \frac{i_C}{i_B} \qquad (2\text{-}1)$$

The ratio h_{FE} is approximately independent of the supply V_{CC} and the load. Example (2-1) shows how we can calculate a typical value of h_{FE}.

Example 2-1: Calculate h_{FE} when $i_C = 1$ mA and $i_B = 10$ μA.
Solution:

$$h_{FE} = \frac{i_C}{i_B} = \frac{1 \times 10^{-3}}{10 \times 10^{-6}} = 100$$

Typical values of h_{FE} for a transistor run from 40 to 600.

It is apparent from Fig. 2-2 that the output terminals of transistors act like an approximate or practical current source, since the collector current

18 Chap. 2 / The Transistor as Amplifier

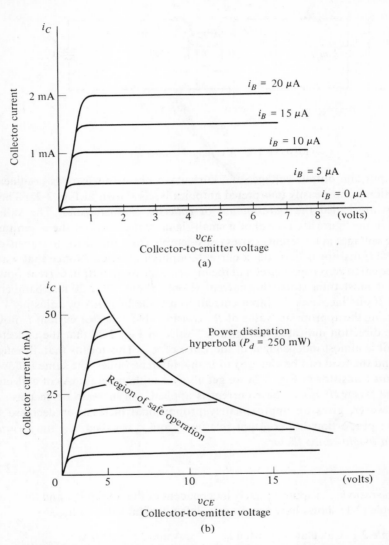

FIGURE 2-2 Transistor static characteristics. (a) Typical output characteristics. (b) Typical output characteristics with maximum power dissipation hyperbola.

i_C is almost independent of v_{CE}. This is true provided the range of operation of the transistor is not exceeded. This range is expressed by the maximum power dissipated by the collector P_d, as shown in Fig. 2-2b. This figure is usually given by the manufacturer for a particular transistor and is usually about 250 milliwatts (mW) for common small-signal transistors and much higher for power transistors.

This maximum power dissipation means that the product $v_{CE} \cdot i_C$ should always be less than P_d, that is,

$$v_{CE} i_C < P_d \tag{2-2}$$

The region of operation of a transistor can be sketched on the static collector characteristics and is then bounded on the v_{CE}-i_C plane by the equation

$$v_{CE} i_C = P_d \tag{2-3}$$

The plot of this equation is called the maximum power dissipation hyperbola. In Fig. 2-2b, this region of safe operation is indicated by the shaded area. v_{CE} and i_C should not exceed these values, or there is danger of destroying the transistor.

2.2 Input Behavior of a Transistor

In Fig. 2-2 we see that we have displayed three of the possible four voltage and current variables for the transistor, v_{CE}, i_C, and i_B. The quantity that we have not yet shown is the input voltage v_{BE}. The static input characteristics are a plot of v_{BE} versus i_B with various values of v_{CE} as the running parameter, as shown in Fig. 2-3. From Fig. 2-3 it is apparent that the base-to-emitter voltage v_{BE} is almost always equal to the value V_{BE}, provided v_{CE} is greater

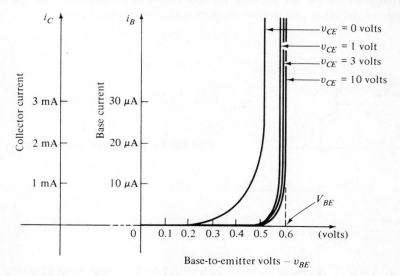

FIGURE 2-3 Static input characteristics for a typical transistor.

than about 1 V and i_B is greater than a few microamperes. For transistors made from silicon, V_{BE} is 0.7 (or 0.6) and 0.3 V for transistors made from germanium. This value does not change significantly if the output voltage v_{CE} is increased or decreased. Since the collector current is approximately proportional to the base current, we need only draw a new scale to obtain i_C, as shown in Fig. 2-3.

2.3 Symbols for a Transistor

So far we have seen the input and output action of the three-terminal device called the transistor. Rather than draw a picture every time we use it in a circuit, we shall use the symbol shown in Fig. 2-4. This type of transistor is called an *npn*. The name *npn* is obtained from the type of materials used in the construction of the transistor. We shall not concern ourselves with this at this time, since it will be covered in a later chapter.

The transistor that complements the *npn* transistor is the *pnp* type. The operation of the *pnp* transistor is identical to the *npn* transistor, except that all the currents flow in the opposite direction to that of the *npn* transistor. That is, to obtain the static input and output characteristics as discussed in Sections 2.1 and 2.2, we merely reverse the polarities of V_{BB} and V_{CC}. The flow of conventional currents in a *pnp* transistor are indicated in Fig. 2-5. Notice that the arrow that designates the emitter is pointing inward in a *pnp* transistor. Since the reference directions for v_{BE} and v_{CE} are as shown, the actual voltages will be negative for *pnp* transistors. Otherwise, the general shape of the input and output characteristics will be the same.

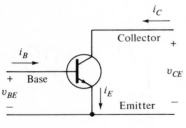

FIGURE 2-4 Symbol for a *npn* transistor.

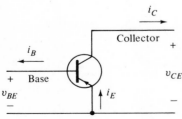

FIGURE 2-5 Symbol for a *pnp* transistor.

If we now refer either to Fig. 2-4 or Fig. 2-5, we use Kirchhoff's current law to write down the relationship

$$i_E = i_B + i_C \tag{2-4}$$

We are really saying in Eq. (2-4) that all the currents entering the transistor must equal all the currents leaving the transistor. Because the collector current is much larger than the base current, the emitter current must approximately equal the collector current. That is, because

$$i_C = h_{FE} i_B \tag{2-5}$$

and h_{FE} ranges from 40 to 600, it follows that since $i_C \gg i_B$,

$$i_E \simeq i_C \tag{2-6}$$

This result can be used as a memory tool to remember the flow of current in the *pnp* and *npn*, as shown in Fig. 2-6. The arrow on the transistor symbol shows the direction of the emitter and collector current, which must flow through the transistor to make it operate correctly. The flow of base current can be remembered by the fact that the emitter current is actually larger than the collector current, since

$$i_E = i_B + i_C \tag{2-7}$$

and the base current must be shown in such a direction so that this is true.

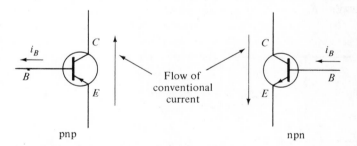

FIGURE 2-6 Flow of conventional current in transistors.

2.4 General Amplifiers

Until now we have looked only at the properties of a device called a transistor. The next logical question one may ask is: What can we use a transistor for? The answer of course is: We use the transistor to amplify weak signal voltages or currents.

The fact is that the study of electronics is centered on the *amplifier*, a device that increases the power level of an input signal. This may be achieved by either increasing the magnitude of the available voltage or current or possibly increasing the magnitude of both. For example, the small amounts of power collected by microphone from sound waves in the air are not nearly enough to drive the voice coil of a loudspeaker for use in an intercom, or to drive the transmitter in a radio or TV station. Amplifiers are used to bring up the level of this input power to a point where it can be used with these devices.

We have already seen that a transistor can probably be used as an amplifier, since the collector current is h_{FE} times as large as the base current. By using the input signal from a device such as a microphone to control the base current, a considerable amplification of current can be obtained, if we consider the collector current to be controlled by the base current.

2.5 Transistor as Current Amplifier

To demonstrate the action of a transistor, we have used two batteries, as shown previously in Fig. 2-1. One battery could be used to combine the function of V_{BB} and V_{CC}. R_s, the source resistance, can now be used to control the base current i_B to the desired level. We shall also use the symbol for the transistor in Fig. 2-1, recognizing that current flow indicates that this is an *npn* transistor. The resulting circuit is shown in Fig. 2-7. The load resistor has been designated as R_L.

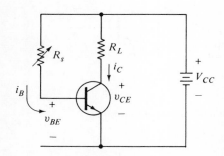

FIGURE 2-7 Transistor in the common emitter configuration.

The value of base current i_B can now be calculated in the following way: The base-to-emitter voltage v_{BE} is approximately 0.6V for a silicon transistor and together with the voltage drop across R_s must add up to V_{CC}, so that we have

$$V_{CC} = i_B R_s + v_{BE} \qquad (2\text{-}8)$$

and

$$i_B = \frac{V_{CC} - v_{BE}}{R_s} \qquad (2\text{-}9)$$

and if
$$V_{CC} \gg v_{BE}$$
then
$$i_B = \frac{V_{CC}}{R_s} \tag{2-10}$$

Using a typical battery voltage of 9 V, we obtain
$$i_B = \frac{8.4}{R_s} \text{ V}/\Omega$$

The value of the collector current is now h_{FE} times larger than i_B, so that
$$i_C = h_{FE} i_B = h_{FE} \frac{8.4}{R_s} \text{ V}/\Omega$$

Notice from the above equation that, by varying R_s, we can vary the collector current. The larger R_s, the smaller the value of i_C.

Example 2-2: Given an *npn* silicon transistor which has a maximum power dissipation of 250 mW, an h_{FE} of 100, a 9-V battery, and a 2-Ω load resistor. The transistor is connected in the manner shown in Fig. 2-8. Calculate the values of R_s so that voltage across the load R_L is (a) 2 V and then (b) 4 V.

Solution:

(a) If the voltage across R_L is to be 2 V, then the collector can be calculated by Ohm's law to be
$$i_C = \frac{v_{RL}}{R_L} = \frac{2 \text{ V}}{2 \text{ k}\Omega} = 1 \text{ mA}$$

Now the base current i_B is h_{FE} times smaller than i_C, so that
$$i_B = \frac{i_C}{h_{FE}} = \frac{1 \text{ mA}}{100} = 10 \text{ }\mu\text{A}$$

The voltage drop across R_s is $v_{RS} = V_{CC} - v_{BE} = 9.0 - 0.6 = 8.4$ V, and R_s can be determined by Ohm's law to be
$$R_s = \frac{v_{RS}}{i_B} = \frac{8.4 \text{ V}}{10 \text{ }\mu\text{A}} = 840 \text{ k}\Omega$$

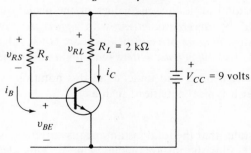

FIGURE 2-8 Circuit for Ex. 2-2.

(b) Using the same approach we can determine the value of R_s necessary for v_{RL} to be 4 V.

$$i_C = \frac{4 \text{ V}}{2 \text{ k}\Omega} = 2 \text{ mA}$$

and

$$i_B = \frac{2 \text{ mA}}{100} = 20 \text{ }\mu\text{A}$$

Hence

$$R_s = \frac{v_{RS}}{i_B} = \frac{8.4 \text{ V}}{20 \text{ }\mu\text{A}} = 420 \text{ k}\Omega$$

From the above example we are able to observe several important results. First, for the typical values used, the value of the base current is larger than 1 or 2 μA, so that v_{BE}, the base-to-emitter voltage approximation of 0.6 V, is very useful. Second, the value of R_s is of the order of about $\frac{1}{2}$ MΩ and is considerably larger than the load resistor R_L. Finally, by varying R_s, which in turn varies the base currrent, we vary the collector current (or the voltage across R_L, which is proportional to i_C).

2.6 Symbols

A system of symbols has been agreed upon by users of transistors and the IEEE, and will be used in the remainder of this book so that the reader will be able to follow the discussions more rapidly. In general, the transistor circuit notation follows a few simple rules:

1. *dc values of quantities* are designated by *capital letters with capital subscripts* (I_B, V_{CE}) and *direct supply voltages* have repeated subscripts (V_{CC}, V_{BB}, V_{EE}).
2. *Small-signal* time-varying components of voltages and currents are represented by *lowercase letters* with *lowercase subscripts* (i_b, i_c).
3. The total instantaneous values or the *actual* quantities are indicated by *lowercase letters* with *uppercase subscripts* (v_{CE}, i_B).
4. Finally, *phasors* or root-mean-square (rms) values of these variations have *uppercase letters* with *lowercase subscripts*.

To help one remember and understand this notation system one should try to remember a typical equation, such as

$$i_B = I_B + i_b \tag{2-11}$$

This equation states that the total instantaneous current i_B is the sum of average (dc value) I_B plus the instantaneous value i_b. This idea is further illustrated in Fig. 2-9.

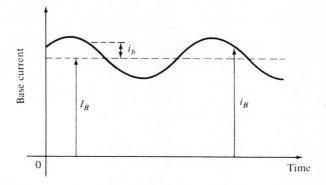

FIGURE 2-9 A diagram of current waveform illustrating the proper use of symbols.

2.7 Transistor as Voltage Amplifier

Before we consider the transistor as a voltage amplifier, consider the action of the circuits shown in Fig. 2-10. The circuits in Figs. 2-10a and 2-10b are identical, except that the circuit in Fig. 2-10b has the voltage v_I separated

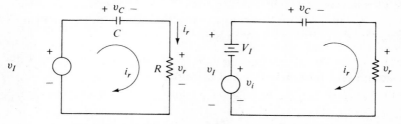

(a) Original circuit   (b) Same circuit with source separated into components.

FIGURE 2-10 A circuit showing the blocking action of a capacitor.

into a dc component V_I and an instantaneous time-varying component v_i, so that

$$v_I = v_i + V_I \tag{2-12}$$

If we apply the fundamental rules from circuit theory, V_C will be equal to V_I, since the capacitor charges up to the value of the battery voltage V_I. Since this is true, no dc current flows through the loop, and i_r, the loop current, is entirely as time-varying signal determined by v_i, C, and R. Using phasor analysis we can determine i_r as follows:

$$I_r = \frac{V_i}{R + j\dfrac{1}{\omega C}} \tag{2-13}$$

and if

$$\frac{1}{\omega C} \ll R$$

then

$$I_r = \frac{V_i}{R} \qquad (2\text{-}14)$$

If the reactance of the capacitor is small in comparison to the value of R, then

$$i_r = \frac{v_i}{R} \qquad (2\text{-}15)$$

or

$$v_i = v_r \qquad (2\text{-}16)$$

Summarizing, we can say that if C is a large capacitor (the reactance is negligible), then the capacitor blocks the flow of dc current and is practically a short circuit to ac.

It is a very similar principle that is used for coupling capacitors in a transistor amplifier, such as the one shown in Fig. 2-11. Let us assume that the source in Fig. 2-11 is a voltage source of a time-varying signal v_i. It delivers a current i_b, which is a time-varying signal with no dc on it, since the capacitor C_c prevents the flow of such a current. Because R_B is a very large resistor, most of the current i_b will flow into the transistor together with current I_B flowing through R_B—to give a total value of i_B, since

$$i_B = I_B + i_b \qquad (2\text{-}17)$$

The voltage across C_c is equal to V_{BE} and is positive, as shown in the diagram. This positive polarity should correspond to positive sign on the capacitor,

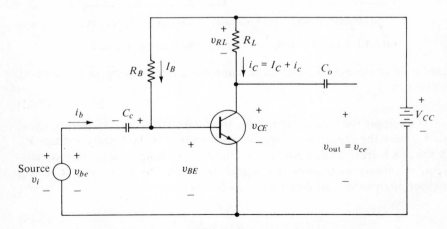

FIGURE 2-11 Voltages and currents in a transistor amplifier.

if an electrolytic capacitor is used. Usually, C_c is about 50 μF for most audio-frequency work. The size of this capacitor will be justified later. We will use 50 μF and say that this size is sufficiently large to be a short circuit for the small-signal current i_b.

Figure 2-12 illustrates the action of the transistor at the input by showing the effect of v_{be} on the input current i_b. The family of static input curves has been approximated by a single curve. The necessity of R_B, which causes I_B to flow into the transistor, is apparent. This dc current I_B fixes the average base current to a point on the linear portion of the curve to produce a current i_b, which has the same shape as v_{be}. I_B is called the *quiescent* or *Q-point* base current. When this is accomplished, we say we have *biased the transistor*. The value of R_B chosen fixes the Q-point current.

Variation in v_{BE} (v_{be}) causes variations in i_B (i_b). These two quantities are related through the slope of the input curve of v_{BE} versus i_B, which is equal

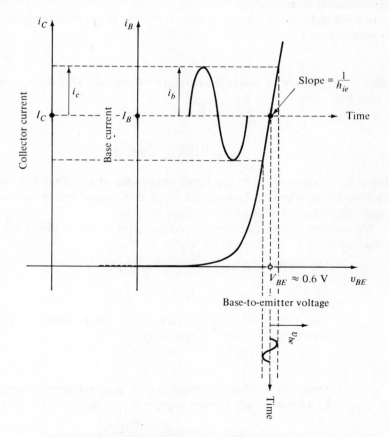

FIGURE 2-12 Input action of a transistor amplifier.

28   Chap. 2 / The Transistor as Amplifier

to $1/h_{ie}$. That is,

$$h_{ie} = \frac{v_{be}}{i_b} \tag{2-18}$$

The variation of base current causes variations in the collector current i_C (i_c). We define a small-signal current gain to be

$$h_{fe} = \frac{i_c}{i_b} \tag{2-19}$$

and is distinguished from h_{FE}, which is given by

$$h_{FE} = \frac{I_C}{I_B} \tag{2-20}$$

the dc current gain. It is apparent that if the lines are equally spaced in the static collector characteristics, the variational and the dc current gains are equal. However, for a practical transistor this is not true and h_{fe} is only approximately equal to h_{FE}.

In solid-state physics it can be shown that h_{ie} is approximately inversely proportional to i_B or

$$h_{ie} \simeq \frac{v_T}{i_B} \tag{2-21}$$

where $v_T \simeq 0.026$ V at room temperature. A typical value for h_{ie} when I_B is 10 μA is

$$\begin{aligned} h_{ie} &\simeq \frac{v_T}{i_B} \simeq \frac{v_T}{I_B} \\ &= \frac{0.026}{10 \times 10^{-6}} = 2.6 \text{ k}\Omega \end{aligned} \tag{2-22}$$

Hence h_{ie} is approximately the input impedance of the amplifier, since it describes the relationship between the input small-signal voltage and current (neglecting the effect of R_B, which is large anyway).

We are now able to calculate the voltage gain of the amplifier according to the following simple steps (refer to Fig. 2-11):

1. The variational base current is given by

$$i_b = \frac{v_{be}}{h_{ie}} = \frac{v_i}{h_{ie}} \tag{2-23}$$

2. The variational collector current i_c is proportional to the base current and is therefore given by

$$i_c = \frac{h_{fe} v_i}{h_{ie}} \tag{2-24}$$

3. Since the collector current has a time-variational component i_c, a time-varying voltage appears across R_L and is given by

$$v_{rl} = i_c R_L = \frac{h_{fe} v_i}{h_{ie}} R_L \tag{2-25}$$

4. Now

$$v_{ce} = -v_{rl} = -\frac{h_{fe}v_i}{h_{ie}} R_L \qquad (2\text{-}26)$$

This result can be reasoned out as follows:

We may say that, since the total voltage v_{CE} and v_{RL} must add up to a constant V_{CC}, v_{CE} and v_{RL} are equal but opposite in sign and so are the variations v_{ce} and v_{rl}.

5. The voltage gain A_v, defined as the ratio output voltage to input voltage, finally gives us, from step 4,

$$A_v = \frac{v_{ce}}{v_i} = -\frac{h_{fe}R_L}{h_{ie}} \qquad (2\text{-}27)$$

The negative sign for the voltage-gain expression means that there is a positively increasing signal at the input and a negatively increasing signal at the output. If the input is a sinusoidal voltage, then the output will be 180° out of phase with the input.

Example 2-3: Given a *pnp* germanium transistor with $h_{FE} = h_{fe} = 40$, a 22 V power supply and a 8.2-kΩ load resistor. The maximum power dissipation of this transistor is 200 mW. See Fig. 2-13.
(a) Bias the transistor at a reasonable Q-point.
(b) Calculate the voltage gain of the amplifier obtained by using the same Q-point values as in (a).

Solution:

(a) The solution to this part of the problem is the same as in Example 2-2 except that this time we have a *pnp* germanium transistor. This means that V_{CC} is negative and that V_{BE} is now -0.3 V. Biasing the transistor at a reasonable Q-point means that V_{CE} should be about one half the value of V_{CC}. This is so that variations that take place in v_{CE} may increase as much as they are able to decrease. This gives a maximum value of voltage swing in v_{CE} at the output, so that the largest variety of magnitudes at the input are acceptable.

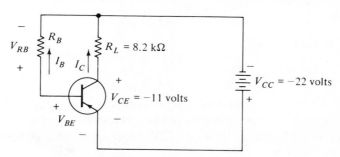

FIGURE 2-13 Figure for Ex. 2-3.

30 Chap. 2 / The Transistor as Amplifier

If V_{CE} is about one half V_{CC}, the same should be true of V_{RL}. Hence the value of I_C, the Q-point collector current, can be determined:

$$I_C = \frac{11 \text{ V}}{8.2 \text{ k}\Omega} = 1.35 \text{ mA}$$

Now the base current and R_B can be calculated:

$$I_B = \frac{I_C}{h_{FE}} = \frac{1.35 \text{ mA}}{40} = 32 \ \mu\text{A}$$

$$R_B = \frac{V_{RB}}{I_B} = \frac{22 - 0.3}{32 \times 10^{-6}} = 710 \text{ k}\Omega$$

Notice that, for biasing, all the symbols are uppercase letters with uppercase subscripts. Now we proceed to make an amplifier from the above circuit.

(b) To design the amplifier we first add in our coupling capacitors, observing the correct polarity, as shown in Fig. 2-14. The voltage gain is given by the expression

$$A_v = -\frac{h_{fe} R_L}{h_{ie}}$$

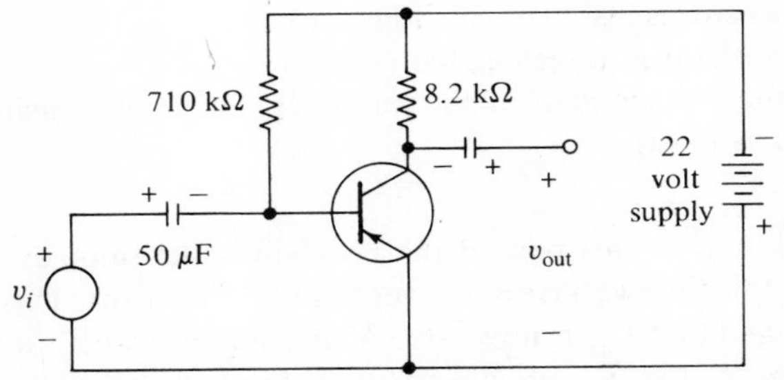

FIGURE 2-14 Circuit for Example problem 2-3.

The quantity h_{ie} can be determined from Eq. (2-21) as follows:

$$h_{ie} = \frac{v_T}{I_B} = \frac{0.026}{32 \times 10^{-6}} = 810 \ \Omega$$

Hence the voltage gain is

$$A_v = -\frac{h_{fe} R_L}{h_{ie}} = -\frac{40 \times 8.2}{810} = -400$$

We see that the voltage gain of a single transistor is quite high. Thus the maximum signal that is used at input must be quite small, since the maximum output peak-to-peak voltage of v_{CE} is 22 V. In this example, the maximum voltage at the input is

$$v_i \text{ (peak to peak)} = \frac{v_{CE}}{|A_v|} = \frac{22}{400} = 55 \text{ mV}$$

If a larger input voltage is used, the output will be distorted, since the collector voltage v_{CE} cannot exceed the supply voltage. This occurs when the variational base current has caused the collector current to reach very near the zero value.

PROBLEMS

2.1 Sketch static output and input curves for a germanium *pnp* transistor similar to those in Figs. 2-2a and 2-2b. The transistor has a dc current gain of 20.

2.2 Sketch the power dissipation hyperbola for the transistor in Problem 2.1 if
  (a) $P_d = 200$ mW (c) $P_d = 1.0$ W
  (b) $P_d = 50$ mW

2.3 (a) If h_{FE} of the transistor shown in Fig. P2-3 is 100, what value of R_B will be needed to make $V_{CE} = 5$ V, as indicated?
  (b) What is the value of h_{ie} for this transistor as biased in this circuit?

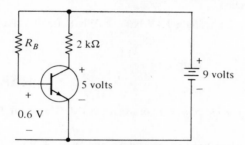

FIGURE P2-3

2.4 Set up a biasing circuit for the transistor in Problem 2.1. Change the value of R_L by 50 per cent and determine the value of V_{CE} now. Is your amplifier still biased? Give a reason for your answer.

2.5 (a) In Fig. P2-5, $I_B = 10$ μA and $h_{fe} = 120$. What will be the voltage gain v_o/v_i for this amplifier? Assume that C_o and C_c are large capacitors.

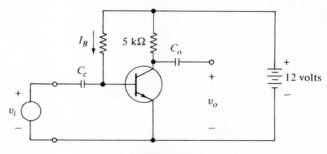

FIGURE P2-5

  (b) If $C_c = 100$ μF and the frequency of $v_i = 40$ Hz, what is the reactance of coupling capacitor C_c? How does this compare with the value of h_{ie}

for this transistor, and can we say that this capacitor is sufficiently large so that it is a short circuit to the current delivered by v_i? Give reasons for your answers.

2.6 For the amplifier shown in Fig. P2-5 with $I_B = 12\ \mu\text{A}$ and $h_{fe} = 130$, what will the maximum peak-to-peak voltage at input be that the circuit can handle before v_o appears distorted on an oscilloscope?
Hint: Assume that $h_{fe} = h_{FE}$ and calculate the gain of the amplifier first.

2.7 (a) For the amplifier shown in Fig. P2-5, why can the value of C_o be quite small (about 0.01 μF) if an oscilloscope is used to measure v_o, even when the frequency of v_i is as low as 20 Hz?
(b) Make up a numerical example showing the type of reasoning you would use to deduce the value of C_o.

2.8 Design an amplifier to have a voltage gain of 500 and input impedance of approximately 2.5 kΩ. A 9-V battery is available to supply the power needed. Describe the type of transistor you will need and give the values of all the components needed.

2.9 Repeat Problem 2.8 using a 12-V battery. What circuit values change in this new design?

REFERENCES

[1] CULTER, P., *Semiconductor Circuit Analysis* (McGraw-Hill Book Co., New York, 1964).

[2] GRAY, P. E., *Introduction to Electronics* (John Wiley & Sons, Inc., New York, 1967).

[3] LEVINE, S. N., *Principles of Solid-State Microelectronics* (Holt, Rinehart and Winston, Inc., New York, 1963).

[4] MATTSON, R. H., *Electronics* (John Wiley & Sons, Inc., New York, 1966).

3

TWO-PORT EQUIVALENT CIRCUITS AND TWO-PORT PARAMETERS

At this point we may feel rather confident about our knowledge of transistor design, since we are able to design and understand the amplifying action of a transistor amplifier. The truth of the matter is that in Chapter 2 we have made many oversimplified approximations and that a better understanding of the transistor may be gained by introducing an electrical equivalent circuit or a model. To construct such a model two approaches are commonly used. First, the electrical properties are treated as mathematical functions (a pair of simultaneous equations) and a model is derived from these functions; then measurements are made to obtain the appropriate constants in the equations. In the second approach, the model is obtained from electronic physics and the dimensions of the transistor.

The former approach is used in this chapter. The constants are obtained from a graphical interpretation of the input and output characterisitics of a transistor.

3.1 Definitions and Terminology for Two-Ports

A *port* is defined as a pair of terminals where a voltage of current may be applied or measured. Thus a *two-port* or *two-port network* is a network with two such sets of terminals. Such a network is shown in Fig. 3-1. The respective ports are usually labeled port 1 and port 2, with appropriate subscripts for voltage and current designating the ports. Two-port theory is thus useful for describing transistors, since a three-terminal device can be thought of as

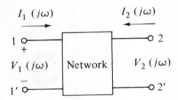

FIGURE 3-1 A two-port network with specified voltage and current reference directions.

having two ports. Of course, for the three-terminal device, the terminals $1'$ and $2'$ shown in Fig. 3-1 are common.

The voltages and currents V_1, V_2, I_1, and I_2 are written as functions of frequency ω but, for resistive networks, may be written in the time domain as v_1, v_2, i_1, and i_2. This is commonly done for transistors operating at audio or low frequencies. Thus the device in the box in Fig. 3-1 can be characterized by the four variables V_1, I_1, V_2, and I_2, which are measurable by external instruments. Of the four variables, only two of these are independent, since at any particular time we can only apply two independent sources to such a network and measure the other two variables. In this case it is sufficient to write only two simultaneous equations to completely characterize the device in the box. For example, one would write

$$V_1 = z_{11}I_1 + z_{12}I_2 \qquad (3\text{-}1)$$

and

$$V_2 = z_{21}I_1 + z_{22}I_2 \qquad (3\text{-}2)$$

The equations are commonly rewritten in matrix form as follows:

$$\begin{bmatrix} V_1 \\ V_2 \end{bmatrix} = \begin{bmatrix} z_{11} & z_{12} \\ z_{21} & z_{22} \end{bmatrix} \begin{bmatrix} I_1 \\ I_2 \end{bmatrix} \qquad (3\text{-}3)$$

Matrix equations are only a shorthand notation for rewriting these equations, and the mathematics of matrix equations may be useful but not necessary to understand the use of two-port theory [1]. The parameters z_{11}, z_{12}, z_{21}, and z_{22} are called *open-circuit impedance parameters* or simply *z-parameters*.

It is common to call the port indicated by the terminals 1 and $1'$ as the input port, and the port indicated by the terminals 2 and $2'$ as the output port. Hence V_1 and I_1 become the input voltage and current, and V_2 and I_2 become the output voltage and current.

3.2 Types of Two-Port Parameters

The *z*-parameters are not the only way to characterize a two-port network. There are five other sets of equations commonly in use for this purpose:

a. **Hybrid or h-Parameters.**

$$\begin{bmatrix} V_1 \\ I_2 \end{bmatrix} = \begin{bmatrix} h_{11} & h_{12} \\ h_{21} & h_{22} \end{bmatrix} \begin{bmatrix} I_1 \\ V_2 \end{bmatrix} \qquad (3\text{-}4)$$

b. **Short-Circuit Admittance or y-Parameters.**

$$\begin{bmatrix} I_1 \\ I_2 \end{bmatrix} = \begin{bmatrix} y_{11} & y_{12} \\ y_{21} & y_{22} \end{bmatrix} \begin{bmatrix} V_1 \\ V_2 \end{bmatrix} \qquad (3\text{-}5)$$

c. **Cascade or $ABCD$ Parameters.**

$$\begin{bmatrix} V_1 \\ I_1 \end{bmatrix} = \begin{bmatrix} A & B \\ C & D \end{bmatrix} \begin{bmatrix} V_2 \\ -I_2 \end{bmatrix} \qquad (3\text{-}6)$$

d. **"Inverse Hybrid" of g-Parameters.**

$$\begin{bmatrix} I_1 \\ V_2 \end{bmatrix} = \begin{bmatrix} g_{11} & g_{12} \\ g_{21} & g_{22} \end{bmatrix} \begin{bmatrix} V_1 \\ I_2 \end{bmatrix} \qquad (3\text{-}7)$$

e. **Inverse Cascade Parameters.**

$$\begin{bmatrix} V_2 \\ I_2 \end{bmatrix} = \begin{bmatrix} b_{11} & b_{12} \\ b_{21} & b_{22} \end{bmatrix} \begin{bmatrix} V_1 \\ -I_1 \end{bmatrix} \qquad (3\text{-}8)$$

For all the above pairs of equations it is possible to obtain an equivalent circuit that can be used for analysis of transistor amplifiers. Although any of these pairs of equations can be used, the z-, h-, and y-parameters are most popular, since they are the easiest to measure. A knowledge of any one set of the parameters, however, is sufficient to work out any of the others. These interrelationships are given in many texts on circuit theory [2, 3].

3.3 Hybrid Equivalent Circuit

1742234

The hybrid equivalent circuit is one of the most popular because of the ease with which it can be measured and used. The pair of hybrid equations can be written as follows:

$$V_1 = h_{11}I_1 + h_{12}V_2 \qquad (3\text{-}9)$$
$$I_2 = h_{21}I_1 + h_{22}V_2 \qquad (3\text{-}10)$$

These parameters can be mathematically deduced as follows and are named by the method by which one measures them:

1. $h_{11} = \dfrac{V_1}{I_1}\bigg|_{V_2=0}$ (the short-circuit input impedance)

2. $h_{21} = \dfrac{I_2}{I_1}\bigg|_{V_2=0}$ (the forward current gain with output short-circuited)

3. $h_{12} = \dfrac{V_1}{V_2}\bigg|_{I_1=0}$ (voltage feedback ratio with the input open-circuited)

4. $h_{22} = \dfrac{I_2}{V_2}\bigg|_{I_1=0}$ (output admittance with the input open-circuited)

This pair of equations and the parameters can be used to describe any electrical circuit. The parameters need only to be measured in the fashion indicated above and then may be used to describe the operation of this network with various devices connected to the ports. However, since most engineers find equivalent circuits easier to handle than systems of equations, an equivalent circuit, as shown in Fig. 3-2, can be quickly deduced. The justification for the equivalent circuit shown in Fig. 3-2 is dual. First, the application of Kirchhoff's voltage law to the portion of the circuit on the left yields the first of the two-port parameter equations,

$$V_1 = I_1 h_{11} + h_{12} V_2 \tag{3-11}$$

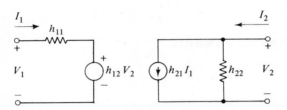

FIGURE 3-2 The h-parameter equivalent circuit for a two-port network.

Second, the application of Kirchhoff's current law to the portion of the circuit on the right leads to the second two-port parameter equation

$$I_2 = h_{21} I_1 + h_{22} V_2 \tag{3-12}$$

Since an equivalent circuit can be established for any network, let us gain some practice by finding the h-parameter equivalent circuit for the simple resistive network given in Fig. 3-3.

Example 3-1: Determine the h-parameter equivalent circuit of the network of the three resistors shown in Fig. 3-3.

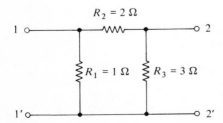

FIGURE 3-3 Resistive two-port for Ex. 3-1.

Solution:

The *h*-parameters can be obtained in the manner shown below:

(a) $$h_{11} = \left.\frac{V_1}{I_1}\right|_{V_2=0}$$

Figure 3-4 indicates the procedure for obtaining h_{11} and h_{21}. From Fig. 3-4,

$$h_{11} = \frac{V_1}{I_1} = \frac{R_1 R_2}{R_1 + R_2} = \frac{2}{3}\,\Omega$$

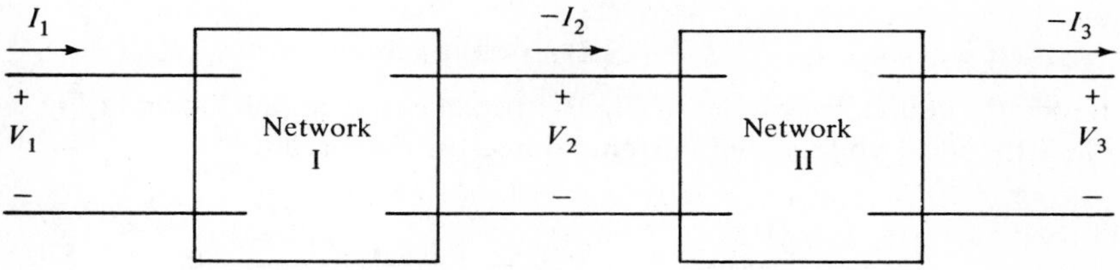

FIGURE 3-4 Resistive two-port with port 2 shorted for Ex. 3-1.

(b) $$h_{21} = \left.\frac{I_2}{I_1}\right|_{V_2=0}$$

Also from Fig. 3-4 and by current division,

$$I_2 = \frac{-\frac{1}{R_2} I_1}{\frac{1}{R_1} + \frac{1}{R_2}}$$

$$= \frac{-\frac{1}{2} I_1}{1 + \frac{1}{2}} = -\frac{1}{3} I_1$$

and

$$h_{21} = -\frac{1}{3}$$

(c) $$h_{12} = \left.\frac{V_1}{V_2}\right|_{I_1=0}$$

Figure 3-5 indicated the procedure for obtaining h_{12} and h_{22}. From Fig. 3-5 and by voltage division,

FIGURE 3-5 Resistive two-port with port 1 open circuit for Ex. 3-1.

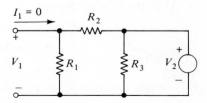

$$V_1 = \frac{R_1}{R_1 + R_2} V_2$$

so that

$$h_{12} = \frac{1}{1+2} = \frac{1}{3}$$

(d)
$$h_{22} = \left.\frac{I_2}{V_2}\right|_{I_1=0}$$

From Fig. 3-5,

$$\frac{1}{h_{22}} = \frac{R_3(R_1 + R_2)}{R_3 + R_1 + R_2}$$

$$h_{22} = \tfrac{2}{3}\,\mho$$

Hence the equivalent circuit of Fig. 3-3 becomes the circuit shown in Fig. 3-6. Note the ideal voltage and current sources in the circuit.

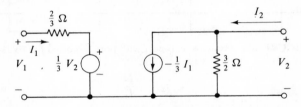

FIGURE 3-6 Final h-parameter equivalent circuit for Ex. 3-1.

In the solution of the above example it is worth noting an issue that may appear quite trivial at present but that will be useful for future problem solving. In Fig. 3-5 and in the solution of h_{12}, the resistor R_3 contributes nothing to the voltage division action. This is true of any resistor in parallel with an ideal voltage when the voltage division formula is used. It is also worth noting that circuits in Figs. 3-6 and 3-3 act identically as far as the terminal properties of the two-port networks are concerned.

In transistor work, the subscripts of the h-parameters have been standardized by the IEEE through the use of two-letter subscripts; the *first* letter indicates the *parameter*, and the *second* the *configuration* in which the measurements were taken. So far, the discussion in this book on transistors has been in the common emitter mode, so that the second subscript used is the lowercase "e."

Thus the four parameters for the common emitter configuration become

$$h_{11} = h_{ie} \text{ (}i\text{ representing } input\text{)}$$

$$h_{12} = h_{re} \text{ (}r\text{ representing } reverse\text{)}$$

$$h_{21} = h_{fe} \text{ (}f\text{ representing } forward\text{)}$$

$$h_{22} = h_{oe} \text{ (}o\text{ representing } output\text{)}$$

Sec. 3.3 / Hybrid Equivalent Circuit 39

In Chapter 2, we discussed the action of the transistor amplifier (in the common emitter configurations) through the static input and output characteristics. These families of curves can be represented mathematically by the pair of functional equations

$$v_{BE} = f(i_B, v_{CE}) \quad \text{input characteristics} \tag{3-13}$$

and

$$i_C = f(i_B, v_{CE}) \quad \text{output characteristics} \tag{3-14}$$

Now the differential change in the input voltage v_{BE} becomes

$$dv_{BE} = \frac{\partial v_{BE}}{\partial i_B} di_B + \frac{\partial v_{BE}}{\partial v_{CE}} dv_{CE} \tag{3-15}$$

and the differential change in the output current becomes

$$di_C = \frac{\partial i_C}{\partial i_B} di_B + \frac{\partial i_C}{\partial v_{CE}} dv_{CE} \tag{3-16}$$

But according to the notation established in Chapter 2 the differential changes can be represented as small-signal voltages and currents as follows:

$$di_C = i_c \quad \text{and} \quad dv_{BE} = v_{be}$$
$$di_B = i_b \quad \text{and} \quad dv_{CE} = v_{ce}$$

Substituting these quantities in Eqs. (3-15) and (3-16), we can rewrite them in the following fashion:

$$v_{be} = \frac{\partial v_{BE}}{\partial i_B} i_b + \frac{\partial v_{BE}}{\partial v_{CE}} v_{ce} \tag{3-17}$$

$$i_c = \frac{\partial i_C}{\partial i_B} i_b + \frac{\partial i_C}{\partial v_{CE}} v_{ce} \tag{3-18}$$

If v_{be} and i_b are taken as the voltage are current at port 1, and v_{ce} and i_c as the voltage and current at port 2, we recognize Eqs. (3-17) and (3-18) to be

$$v_{be} = h_{ie} i_b + h_{re} v_{ce} \tag{3-19}$$

and

$$i_c = h_{fe} i_b + h_{oe} v_{ce} \tag{3-20}$$

where

$$h_{ie} = \frac{\partial v_{BE}}{\partial i_B} \bigg|_{\substack{v_{ce}=0 \\ v_{CE}=\text{constant}}} \quad * \tag{3-21}$$

$$h_{re} = \frac{\partial v_{BE}}{\partial v_{CE}} \bigg|_{\substack{i_b=0 \\ i_B=\text{constant}}} \tag{3-22}$$

$$h_{fe} = \frac{\partial i_C}{\partial i_B} \bigg|_{\substack{v_{ce}=0 \\ v_{CE}=\text{constant}}} \tag{3-23}$$

$$h_{oe} = \frac{\partial i_C}{\partial v_{CE}} \bigg|_{\substack{i_b=0 \\ i_B=\text{constant}}} \tag{3-24}$$

*The definition of h_{fe} and h_{ie} is given here in more precise and rigorous terms than in Chapter 2.

Thus the transistor, for differential changes in voltages and currents, can be represented by the circuit shown in Fig. 3-7a, provided the transistor is in the common emitter configuration, as indicated in Fig. 3-7b.

The preceding derivation shows in differential form the method for obtaining the h-parameters of a transistor biased in the common emitter configuration. The equivalent circuit in Fig. 3-7a thus shows the action of a transistor for variations of voltages and currents occurring about a constant or Q-point value, as biased by the designer.

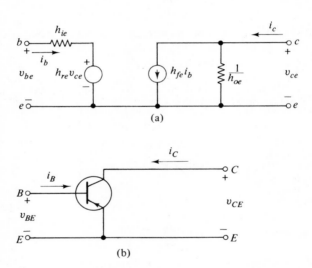

(a) h-parameter equivalent circuit of a transistor in the common emitter configuration.

(b) Reference directions of voltages and currents for a transistor in the common emitter configuration.

FIGURE 3-7 The common emitter transistor configuration.

3.4 Characteristic Curves and h-Parameters

The h-parameters, as derived in Section 3.3 in differential form, can be approximated by incremental changes. That is,

$$h_{ie} = \left.\frac{\partial v_{BE}}{\partial i_B}\right|_{v_{CE}=\text{constant}} \simeq \left.\frac{\Delta v_{BE}}{\Delta i_B}\right|_{v_{CE}=V_{CE}} \tag{3-25}$$

$$h_{re} = \frac{\partial v_{BE}}{\partial v_{CE}}\bigg|_{i_B=\text{constant}} \simeq \frac{\Delta v_{BE}}{\Delta v_{CE}}\bigg|_{i_B=I_B} \quad (3\text{-}26)$$

$$h_{fe} = \frac{\partial i_C}{\partial i_B}\bigg|_{v_{CE}=\text{constant}} \simeq \frac{\Delta i_C}{\Delta i_B}\bigg|_{v_{CE}=V_{CE}} \quad (3\text{-}27)$$

$$h_{oe} = \frac{\partial i_C}{\partial v_{CE}}\bigg|_{i_B=\text{constant}} \simeq \frac{\Delta i_C}{\Delta v_{CE}}\bigg|_{i_B=I_B} \quad (3\text{-}28)$$

The parameters h_{fe} and h_{ce} are obtained from the static output characteristics, as shown in Fig. 3-8. The parameter h_{fe} has been defined previously and is the ratio of the change in the collector current and the change of the base current at a fixed value of v_{CE}, that is, V_{CE}. The parameter h_{oe} becomes the *slope* of one of the output characteristic curves for a fixed value of I_B.

The parameters h_{ie} and h_{re} are obtained from the static input characteristics in much the same way. The procedure is shown in Fig. 3-9. Notice that the parameter h_{ie} is the *reciprocal* of the *slope* of the input characteristic at a fixed value of v_{CE}, namely V_{CE}. This was demonstrated in Chapter 2. The parameter h_{re} is the ratio of the change in the base-to-emitter voltage and the collector-to-emitter voltage for a fixed value of base current, that is, I_B.

To obtain good results when measuring the *h*-parameters from these curves, the increments should be taken as small as possible to ensure good accuracy. Of course, if they are too small, it becomes difficult to read the values from the graph. Good judgment should be exercised in doing this, since the parameters do change at different *Q*-points on the graphs. When one is using these parameters, all the measurements should be taken at *Q*-point of the transistor amplifer, considered in a particular circuit.

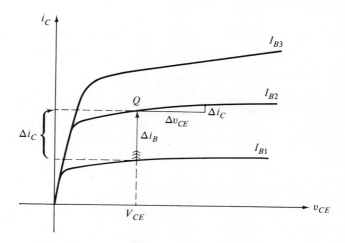

FIGURE 3-8 Illustration of methods for finding h_{fe} and h_{oe} from the output characteristics.

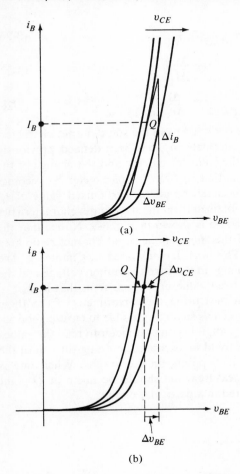

FIGURE 3-9 Illustration of method for finding (a) h_{ie}, (b) h_{re}.

Typical values for h-parameters are given in Table 3-1. It should also be noted that these are relatively low-frequency parameters; for high frequencies different methods are devised for obtaining these parameters. These parameters can generally be used for most low-frequency and audio-amplifier applications when measured from the static characteristics, as discussed in this section.

TABLE 3-1 TYPICAL h-PARAMETER VALUES

Parameter	Typical Values
h_{ie}	2.6×10^3 Ω
h_{fe}	100
h_{re}	2×10^{-4}
h_{oe}	5×10^{-5} mho

3.5 Dynamic Measurement of h-Parameters

One of the first questions asked about the h-parameter equivalent circuit is: How do we know that h_{ie} and h_{oe} are resistors and that h_{fe} and h_{re} are simple ratios? The answer, of course, is that this is true for a limited range of low frequencies. We shall now explain the reasoning for this.

Consider a nonlinear* resistor R_n, which has the arbitrary volt–ampere characteristics shown in Fig. 3-10. Suppose now that a battery of value V_Q is connected to the resistor; then the current that will flow through R_n will be I_Q. To obtain the curve in Fig. 3-10 (the static characteristics), various values of V_Q are applied, and I_Q is measured each time that a different V_Q is applied. The results are then plotted on a graph. Alternatively, one can obtain the characteristics by adding a voltage Δv_R in series with V_Q and then changing the value of Δv_R as shown in Fig. 3-11. In Fig. 3-10, the value of V_Q has been changed to V_B and V_A, with I_B and I_A flowing for each respective value. The question that now remains is: How quickly can one make the meas-

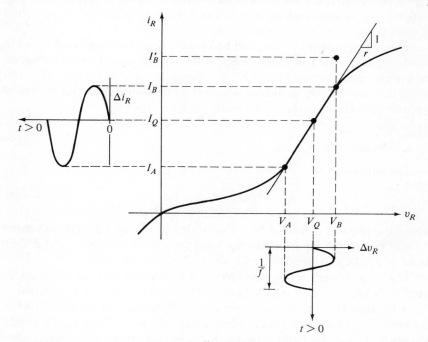

FIGURE 3-10 The v—i curves of a nonlinear and the effect of varying an applied voltage about V_Q.

*A nonlinear resistor is one that does not obey Ohm's law for values of v and i for which the resistor is used.

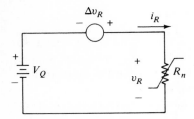

FIGURE 3-11 A circuit showing a nonlinear resistor R_n in dynamic operation.

urement or how quickly can one go from V_Q to V_B and obtain I_B? In a pure resistor the frequency of the measurement is not important, since a resistor always obeys Ohm's law. Hence a resistor such as R_n can be characterized by the slope of the curve in the neighborhood of the Q-point current I_Q, provided change in the voltage from V_Q to V_B produces I_B and not I'_B. If I'_B is produced at a certain rate of measurement, the device is not purely resistive at this frequency, and probably some capacitance should be added in parallel with R_n to account for the increase in current at higher frequencies. In Fig. 3-10, the rate of measurement is shown as f, which is reciprocal of the period of time taken to do the measurement.

It is now apparent that Δv_R can be sinusoidal, and the results obtained in measuring a small-signal characteristic will be the same as the result obtained by a static measurement, provided the frequency is low enough. It is this result that we will use to discuss practical methods for the measurement of h-parameters without tracing the static characteristics.

a. Measurement of h_{ie}.

The parameter h_{ie} is defined as follows:

$$h_{ie} = \frac{\partial v_{BE}}{\partial i_B}\bigg|_{v_{CE}=V_{CE}} = \frac{v_{be}}{i_b}\bigg|_{v_{ce}=0} \tag{3-29}$$

where the Q-point is defined by a fixed value of V_{CE} and I_C. The parameter h_{ie} is of course the input impedance with the output voltage v_{CE} at a fixed value V_{CE}. The measurement technique is shown in Fig. 3-12.

The input impedance is measured in the standard way by adjusting R_s until v_{be} is one half the magnitude of v_i. Under this condition R_s is equal to h_{ie}. A large choke L can be used in series with R_B to increase the impedance of the biasing network and hence improve the accuracy of the measurement. However, L may be omitted, since R_B is usually many times larger than h_{ie}, and error encountered due to R_B is small.

Good accuracy is obtained by measuring v_{be} and v_i with a good ac vacuum-tube voltmeter (VTVM).

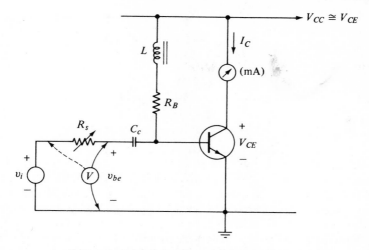

FIGURE 3-12 The circuit for measuring h_{ie}.

b. Measurement of h_{fe}.

The parameter h_{fe} is defined as follows:

$$h_{fe} = \frac{\partial i_C}{\partial i_B}\bigg|_{v_{CE}=V_{CE}} = \frac{i_c}{i_b}\bigg|_{v_{ce}=0} \qquad (3\text{-}30)$$

where the Q-point is also defined by a fixed value of V_{CE} and I_C.

The parameter h_{fe} is the short-circuit current gain of a transistor. The term *short circuit* implies that $v_{ce} = 0$ or that v_{CE} remains fixed. This can only be done in an approximated way, since measurement of the short-circuit current necessitates some variation in v_{CE}. The measurement scheme is shown in Fig. 3-13. The Q-point current I_C is measured by measuring the dc voltage across R_L. To make v_{CE} approximately constant during measurement, R_L is chosen as small as is consistent with measurable voltages. A 100-Ω load resistor is a good compromise. The short-circuit current i_c is measured by measuring the ac component of v_{CE}, which is equal and opposite to that of the voltage across R_L. Thus the short-circuit collector current is

$$i_c = -\frac{v_{ce}}{R_L} \qquad (3\text{-}31)$$

The input current i_b is obtained by measurement of an ac voltage drop across R_s. If an instrument with a common ground with the signal generator is used, then v_i and v_{be} both need to be measured, so that the input current is given by

$$i_b = \frac{v_i - v_{be}}{R_s} \qquad (3\text{-}32)$$

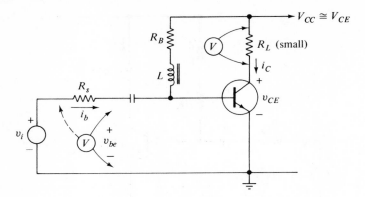

FIGURE 3-13 The circuit for measuring h_{fe}.

To obtain good results, the value of R_s should be of the same order of magnitude or much larger than h_{ie}. This makes the measurement of voltage across R_s quite easy and with not too much error.

c. Measurement of h_{re}.

The parameter h_{re} is defined as follows:

$$h_{re} = \frac{\partial v_{BE}}{\partial v_{CE}}\bigg|_{i_B=I_B} = \frac{v_{be}}{v_{ce}}\bigg|_{i_b=0} \tag{3-33}$$

The measurement technique for h_{re} is shown in Fig. 3-14. In this measurement, h_{re} is found by measuring v_{be} and v_{ce} with an ac VTVM and then taking the appropriate ratio. The constant bias current I_B is provided by R_B and a large inductor L. The variations in v_{CE} are provided by an audio interstage transformer,* which also provides isolation between the ground from the signal generator and the instrument used to measure v_{CE} and v_{be}.

d. Measurement of h_{oe}.

The measurement of h_{oe} is similar to that of h_{re} and h_{fe}. The parameter h_{oe} is defined as

$$h_{oe} = \frac{\partial i_C}{\partial v_{CE}}\bigg|_{i_B=I_B} = \frac{i_c}{v_{ce}}\bigg|_{i_b=0} \tag{3-34}$$

The measurement scheme is shown in Fig. 3-15. The Q-point collector current is measured by measuring the dc voltage across R_L. R_L may be any convenient value that will give good voltmeter readings.

The small-signal collector current i_c is measured by using an ac VTVM across R_L, and the transformer again provides isolation for the ground on the

---

*The action and design of transformers for electronic circuits will be discussed in a later chapter.

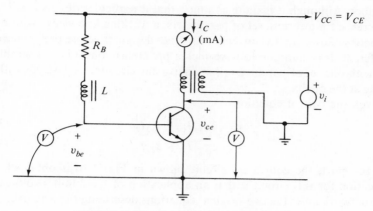

FIGURE 3-14 The circuit for measuring h_{re}.

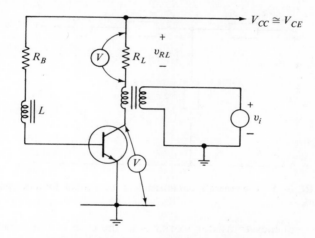

FIGURE 3-15 The circuit for measuring h_{oe}.

voltmeter if there is one. The collector current is obtained by Ohm's law from a knowledge of R_L. R_B and L supply a constant base current, I_B, as required in the definition. The ratio of i_c to v_{ce} (measured with an ac VTVM) gives us h_{oe}.

3.6 z-Parameters and the Equivalent Circuit

Although all the results needed for successful circuit design could be obtained through one set of parameters, such as the *h*-parameters, it is useful to study other parameters and their equivalent circuits. A study of other parameters often leads to a better understanding and in some cases easier formulas to remember. Although it is quite common that a particular designer will have a preference for a certain set of parameters, a working knowledge with other parameters allows him to converse with other designers whose preference may be different. It is common to have such a preference, but it is undesirable to work with one set of parameters (or equivalent circuits) exclusively without looking at the others.

Given the pair of equations

$$V_1 = z_{11}I_1 + z_{12}I_2 \qquad (3\text{-}35)$$

$$V_2 = z_{21}I_1 + z_{22}I_2 \qquad (3\text{-}36)$$

for a two-port, the equivalent circuit shown in Fig. 3-16 is obtained. The justification for this circuit is that an application of Kirchhoff's voltage law to the network gives the two original equations describing the network.

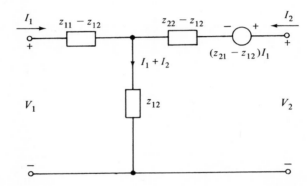

FIGURE 3-16 The general z-parameter equivalent circuit for a two-port network.

For low-frequency transistor work, as in the case of the h-parameters, the impedances shown as boxes in Fig. 3-16 can be replaced by resistors. Also, since we are dealing with small-signal time-varying quantities, the voltages and currents can be written with lowercase letters to denote this. This means that the model will be used about a Q-point, in which the variations from the Q-point will not be too large. This resultant equivalent circuit is shown in Fig. 3-17. The symbolism in Fig. 3-17 is left in a general form, showing input and output quantities with the subscripts i and o, respectively. The circuit shown in Fig. 3-17 can also be converted to have a dependent current generator rather than a dependent voltage generator. This can be done by using Norton's theorem on the resistor r_3 and the voltage source $r_4 i_i$. The resulting circuit is shown in Fig. 3-18. We shall leave these z-parameter circuits for awhile and practice with the h-parameter equivalent circuit first.

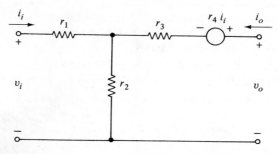

FIGURE 3-17 The z-parameter equivalent circuit for transistors at low frequencies or resistive networks.

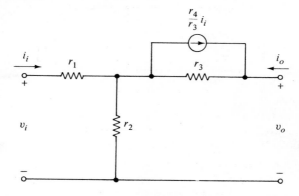

FIGURE 3-18 Low frequency z-parameter equivalent circuit with a dependent current generator.

PROBLEMS

3.1 (a) Obtain the h-parameters for the circuit shown in Fig. P3-1.
(b) What is the h-parameter equivalent circuit for this circuit?

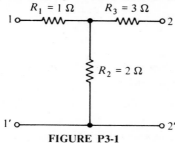

FIGURE P3-1

3.2 Connect an ideal voltage source V_1 to port 1 and a resistor $R_4 = 4\,\Omega$ to port 2. Using the equivalent circuit obtained in Problem 3.1, obtain the voltage V_2 across R_4. Check this result by connecting R_4 in the original network shown in Fig. P3-1.

3.3 (a) Give the mathematical definitions of the $ABCD$ parameters.
(b) Obtain the $ABCD$ parameters of the circuit shown in Fig. P3-1.

3.4 If, in Fig. P3-4, the cascade matrix parameters of Network I are $A_1, B_1, C_1,$ and D_1 and of Network II $A_2, B_2, C_2,$ and D_2, what are the cascade parameters of the network composed of both Network I and Network II?

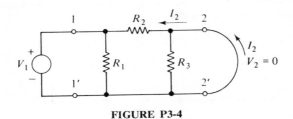

FIGURE P3-4

3.5 If P_d of the transistor is 250 mW, h_{fe} is 100 for all values of i_C at V_{CE} equal to 2 V, and h_{oe} is 5×10^{-5} mho, sketch the static output characters of a transistor in the common emitter configuration.

3.6 Obtain the h-parameter equivalent circuit of the network shown in Fig. P3-6.

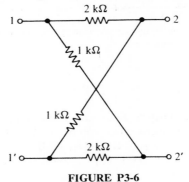

FIGURE P3-6

3.7 Devise techniques, similar to those used to measure h-parameters, to measure the z-parameters of a transistor.

3.8 Repeat Problem 3.7, devising techniques for obtaining y-parameters of a transistor.

3.9 Obtain an equivalent circuit for the circuit in Fig. P3-1 by first obtaining the z-parameters and then using these values in the circuit shown in Fig. 3-16.

REFERENCES

[1] Van Valkenburg, M. E., *Network Analysis*, 2nd ed. (Prentice-Hall, Inc., Englewood Cliffs, N.J., 1964).

[2] Javid, M., and Brenner, E., *Analysis, Transmission, and Filtering of Signals* (McGraw-Hill Book Co., New York, 1963).

[3] Zeines, B., *Introduction to Network Analysis* (Prentice-Hall, Inc., Englewood Cliffs, N.J., 1967).

[4] Middendorf, W. H., *Introductory Network Analysis* (Allyn and Bacon, Inc., Boston, 1965).

[5] General Electric Company, eds., *Transistor Manual*, 7th ed. (General Electric Company, Syracuse, N.Y., 1964).

4

BIASED COMMON EMITTER AMPLIFIER

We have already seen that transistors do not amplify signals by merely connecting a signal source at one end and a load on the other end. The transistor must be biased by means of a battery that not only supplies the energy to achieve amplification but supplies a constant collector and base current, which is necessary to place these currents in favorable operating regions.

Thus far we have only a single method of achieving a biased state and a rather intuitive method for explaining amplification. By use of equivalent circuits and an additional study into biasing we shall be able to understand a larger class of amplifier circuits, other than the simple one discussed in Chapter 2.

4.1 Constant Current Biasing

The method of biasing discussed in Chapter 2 in which two resistors R_B and R_L and a single power supply are used will be termed *constant current biasing*.* This circuit is shown again in Fig. 4-1. The name given to this type of biasing is derived from the fact that the base current, I_B, is fixed and depends on R_B and the supply voltage V_{CC}; the expression for base current is

$$I_B = \frac{V_{CC} - V_{BE}}{R_B} \simeq \frac{V_{CC}}{R_B}$$

*Some texts refer to this configuration as *base bias*.

Sec. 4.1 / Constant Current Biasing 53

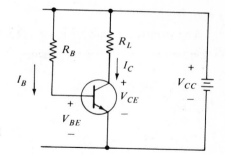

FIGURE 4-1 The constant-current biased transistor.

Since V_{BE} is relatively constant and also negligible for large values of V_{CC}, the base current remains fixed regardless of the transistor used (providing of course it is *npn*, as in Fig. 4-1).

One of the basic problems encountered in the design of transistor amplifiers is that of establishing the proper dc emitter (or collector) current. The biasing problem is due primarily to the change of h_{FE} with temperature, as shown in Fig. 4-2. Notice that, under a change of temperature from 20 to 80°C, h_{FE} may more than double, causing V_{CE} to go down to almost zero, thus preventing negative variations in v_{CE} during operation as an amplifier. This problem is easily demonstrated by means of an example.

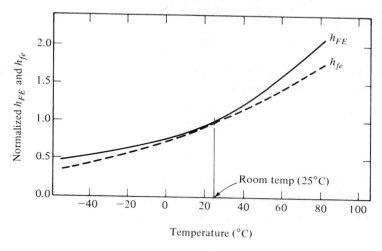

FIGURE 4-2 Typical variations of h_{FE} and h_{fe} with temperature.

Example 4-1: Given a dc supply of 6 V, a load resistor of 1 kΩ, and a germanium *npn* transistor with $h_{FE} = 50$ at room temperature,
(a) Bias the transistor at a reasonable value of V_{CE}.
(b) Calculate the value of V_{CE} if $h_{FE} = 100$ at 70°C. Is the amplifier still biased?

Solution:

The required circuit with a 1-kΩ load resistor is shown in Fig. 4-3.

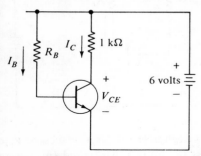

FIGURE 4-3 Transistor circuit for Ex. 4-1.

(a) By KVL (Kirchhoff's voltage law)

$$V_{CC} = V_{CE} + I_C R_L$$

By setting V_{CE} equal to 3 V, we can solve for the collector current:

$$I_C = \frac{V_{CC} - V_{CE}}{R_L} = \frac{6-3}{1 \times 10^3} = 3 \text{ mA}$$

and

$$I_B = \frac{I_C}{h_{FE}} = \frac{3 \text{ mA}}{50} = 60 \text{ } \mu\text{A}$$

$$R_B = \frac{V_{CC} - V_{BE}}{I_B} = \frac{5.7 \text{ V}}{60 \mu\text{A}} = 950 \text{ k}\Omega$$

(b) Since I_B is fixed and h_{FE} has increased to 100, the collector current at 70°C can be obtained as follows:

$$I_C = I_B h_{FE} = 100 \times 60 \mu\text{A}$$
$$= 6 \text{ mA at } 70°C$$

and

$$V_{CE} = V_{CC} - I_C R_L$$
$$= 6.0 - 6.0 = 0.0 \text{ V}$$

This circuit is not biased, since the collector-to-emitter voltage is unable to go down for negative variations in v_{CE}' when this circuit is used as an amplifier.

We can now draw the conclusion that the constant current biasing is not too satisfactory when other considerations (such as temperature, aging, and replacement of transistors) are taken into account. A biasing scheme in which the collector current I_C and the collector-to-emitter voltage V_{CE} remains fixed regardless of the value of h_{FE} is therefore needed to have stable biasing conditions.

4.2 H-type Biasing

In answer to the requirement of improving upon constant current biasing, *H-type* (or sometimes called chain- or emitter-type) biasing has been devised. In most applications, the changes in bias conditions due to a change in h_{FE}, which, in turn, is due to aging or temperature changes, are not acceptable. Thus a biasing design procedure independent of changes in h_{FE} and almost independent of the value of h_{FE} itself would be of great value. This would mean that a designer would be able to bias circuits without knowing the manufacturer's specification of the particular transistor he has, or the value of h_{FE}. In practice, a transistor manufacturer's number would not be of much value anyway, since the value of h_{FE} differs widely from unit to unit. Usually the manufacturer specifies only a range of values for the *h*-parameters.

The circuit called the *H-type biased amplifier stage* is shown in Fig. 4-4.

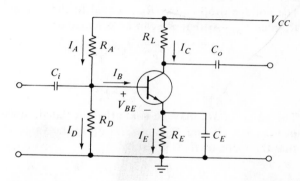

FIGURE 4-4 An *H*-type biased common emitter amplifier.

The design procedure is given in the few simple steps that follow:

1. As in constant current biasing, a suitable collector current I_C is chosen (i.e., a value from 0.2 to 10 mA is used).
2. The value of V_{CC}, the supply, is usually specified for most designs, but if not, its value may now be chosen.
3. Values of R_L and R_E are chosen next. The same values of R_L can be used as in current biasing, and R_E may be the same value but is usually chosen from 2 to 20 times smaller in magnitude than R_L. (A little practice will help the designer to choose reasonable values.)
4. Because $I_C \cong I_E$, the designer can now calculate the voltage drop across R_L and R_E, using the chosen collector or emitter current. V_{CE} may also be deduced, that is,

$$V_{CE} = V_{CC} - I_C R_L - I_E R_E \tag{4-1}$$

As before, V_{CE} should be large enough to allow variations in v_{CE} to occur when a signal is to be applied.

5. A rough estimate of the base current is made, that is,

$$I_B = \frac{I_C}{h_{FE}} \qquad (4\text{-}2)$$

Only a very approximate value is needed here. Then the current I_A is chosen to be at least 10 times that of I_B.

6. Since

$$I_A = I_D + I_B \qquad (4\text{-}3)$$

and

$$I_B < I_A \qquad (4\text{-}4)$$

then

$$I_A \simeq I_D \qquad (4\text{-}5)$$

The voltage drop across R_D is equal to the voltage drop across R_E plus V_{BE}, so that R_D can be calculated:

$$R_D = \frac{I_E R_E + V_{BE}}{I_D} \qquad (4\text{-}6)$$

where

$$I_D \simeq I_A$$

7. The value of R_B may be similarly calculated, since the voltage drop across R_B and R_A must add up to the value V_{CC}. It follows that the value of R_A is given by

$$R_A = \frac{V_{CC} - I_D R_D}{I_A} \simeq \frac{V_{CC} - I_A R_D}{I_A} \qquad (4\text{-}7)$$

8. Finally, a large value of C_E is used as a *bypass capacitor* for R_E. The value of C_E should be such that the value of its reactance at the lowest frequency to be used should be approximately equal to or smaller than h_{ie}/h_{fe}. The reason for this choice will be justified in a later chapter.

Let us now see why this design is temperature stable. The answer lies in the fact that I_A is chosen much larger than the value of I_B. The two resistors R_A and R_D act like a simple two-resistance divider and, since the current I_B is much less than I_A or I_D, the voltage drop across R_D remains fixed even if the temperature changes the value of I_B. Since the voltage across R_D remains fixed, the voltage V_{BE} plus the drop across R_E remains fixed, since they are equal to one another. Now if the temperature changes in V_{BE} are negligible (in many cases this is true), then the emitter current is fixed as well as being approximately equal to the collector current. Since these currents are fixed, so are the voltages across R_L and R_E fixed, as well as V_{CE}.

It is important to notice that the choice of I_A in step 5 of the design procedure determines the temperature stability of the circuit. The larger the ratio of I_A to I_B selected, the more stable the circuit. I_A should not be chosen so large as to make R_D too small, since little is gained when I_A is more than 10 or 20 times larger than I_B. To illustrate the design procedure let us do an example in the same manner as was outlined.

Example 4-2: The circuit in Fig. 4-4 is used and it will be assumed that h_{FE} of the transistor shown is at least 50, since this is the minimum value specified by the manufacturer. Design an H-type biased CE amplifier.

Solution:

(a) We will choose a collector current first. Let

$$I_C = 1 \text{ mA}$$

(b) Choosing the supply voltage, we let

$$V_{CC} = 20 \text{ V}$$

(c) We choose R_L and R_E as

$$R_L = 10 \text{ k}\Omega$$

and

$$R_E = 1 \text{ k}\Omega$$

(d) The voltage drops across R_L and R_E are now calculated:

$$I_C R_L = 1 \text{ mA} \times 10 \text{ k}\Omega = 10 \text{ V}$$

and

$$I_E R_E \cong 1 \text{ mA} \times 1 \text{ k}\Omega = 1 \text{ V}$$

so that

$$V_{CE} \cong V_{CC} - I_C R_L - I_E R_E$$
$$\cong 20 - 10 - 1 = 9 \text{ V}$$

(e) If we choose $I_A \simeq \frac{1}{2}$ mA, then I_A will certainly be larger than 10 times the value of I_B, even if we use the lowest possible value of h_{FE} to get the largest value of I_B. That is,

$$I_{B(\text{maximum})} = \frac{I_C}{h_{FE(\text{minimum})}} = \frac{1 \text{ mA}}{50}$$

(f) Now R_D can be calculated with Eq. (4-6):

$$R_D = \frac{1.6 \text{ V}}{I_D} = \frac{1.6 \text{ V}}{\frac{1}{2} \text{ mA}} = 3.2 \text{ k}\Omega$$

(g) The resistor R_A as calculated by Eq. (4-7) becomes

$$R_A = \frac{(20 - 1.6)\text{V}}{\frac{1}{2}\text{mA}} = 37\text{ k}\Omega$$

(h) Now, to deduce a suitable value of C_E, we can say that

$$\frac{1}{\omega C_E} < \frac{h_{ie}}{h_{fe}}$$

To ensure for most of the audio frequency, a large value such as a 100 μF electrolytic capacitor is chosen of at least 2 WVDC.* The dc voltage across it will be only 1 V, since this is the voltage across R_E.

It is recommended that the reader at this point do a few practical examples to ensure a knowledge of this procedure. Figure 4-4 is probably the most used biasing circuit for amplifiers. A designer must be able to make decisions regarding the values of I_C, V_{CC}, R_L, and R_E to obtain a reasonable value of V_{CE}. There are no set rules, and the designer may wish to use the manufacturer's recommended values for the Q-point (which is V_{CE} and I_C) but this is not necessary at all, as long as we are within the maximum values discussed in Chapter 2.

4.3 Current-Feedback Biasing

Still another feedback technique for bias stabilization can be employed. The circuit for this technique is shown in Fig. 4-5. This current-feedback method provides almost the same stability as the H-type biasing but is not quite as popular because the gain is reduced by feedback provided by R_B.

The stabilizing action of current-feedback biasing can be easily understood in a nonmathematical way by the following steps. An increase in temperature results in an increase in h_{FE}. This corresponds to an increase in I_C for a given I_B. This decreases V_{CE} and thus the voltage across R_B, which in turn decreases I_B. A decrease in I_B now means a decrease in I_C. This is, of course, the usual action of negative feedback.

For the circuit shown in Fig. 4-5, the design procedure is quite straightforward:

1. Assume a collector current that is approximately equal to the current through R_L, since $I_B \ll I_C$.
2. Calculate the value of V_{CE}, choosing a suitable value of R_L.
3. The voltage across R_B, namely V_{RB} can be calculated from

$$V_{RB} = V_{CE} - V_{BE} \tag{4-8}$$

$$R_B = \frac{V_{CE} - V_{BE}}{I_B} \tag{4-9}$$

*The abbreviation WVDC stands for workable volts direct current and refers to the maximum dc voltage that should be applied across the capacitor.

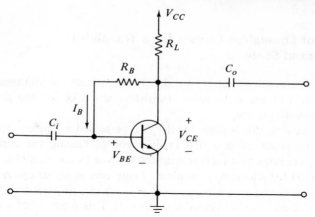

FIGURE 4-5 Current feedback biased common emitter amplifier.

where

$$I_B = \frac{I_C}{h_{FE}} \tag{4-10}$$

As in current-source-type biasing, a knowledge of h_{FE} is necessary for proper biasing for the above. This is perhaps another reason why this biasing technique is not too popular among designers.

Other biasing schemes are sometimes employed. These are usually combinations of the three basic methods just discussed. Two of these are shown in Fig. 4-6. These are quite easy to bias after the three basic methods have been learned and thoroughly understood.

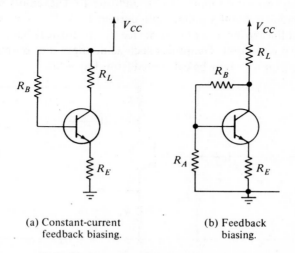

(a) Constant-current feedback biasing.

(b) Feedback biasing.

FIGURE 4-6 Alternate forms of biasing.

4.4 Use of Equivalent Circuit for a Transistor in Biased State

Since the h-parameter model is valid only for variational voltages about the Q-point, we can use it to solve transistor networks for the time-varying component voltages only.

Let us see how this is done for the H-type biased amplifier shown in Fig. 4-7. The first observation is that in this type of biasing the capacitors C_i, C_E, and C_o are large and their reactances are low (a few ohms) in the ranges of frequencies for which v_g is applied. Therefore, in an *ac equivalent circuit* these capacitors can be approximated as short circuits. Second, the battery is a dc source and has no variations across it. This means that it also can be replaced by a short circuit, a zero ac source. The resultant circuit with these modifications is shown in Fig. 4-8a.

Two steps now remain to obtain the equivalent circuit for Fig. 4-7. The first step is to recognize that Node A shown in Fig. 4-8a is really the ground node, and second that the transistor is replaced by its equivalent h-parameter circuit. The resulting circuit is shown in Fig. 4-8b. This figure is called the *ac equivalent circuit* for the amplifier shown in Fig. 4-7.

A *warning* should be given here that the amplifier shown in Fig. 4-8a is not a functional circuit and, strictly speaking, is not even a valid circuit to draw. Its purpose is only to indicate an intermediate step in obtaining the ac equivalent circuit.

The resulting circuit in Fig. 4-8b is now a very useful circuit diagram, because we can now use the theorems in the circuit that apply to linear circuits. We can now use Ohm's law in addition to Thévenin's and Norton's theorems and Kirchhoff's voltage and current laws. We have succeeded, in Fig. 4-8b, in linearizing a nonlinear device (the transistor) but only for variations about the Q-point. Quantities such as ac input and output impedance and voltage gain may now be calculated from Fig. 4-8b.

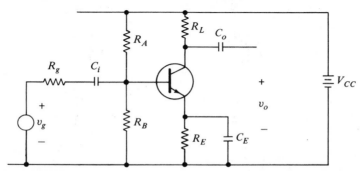

FIGURE 4-7 Complete h-type biased common emitter amplifier circuit.

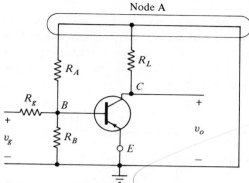

(a) The first step in obtaining the ac equivalent circuit by replacing all large capacitors and batteries with short circuits

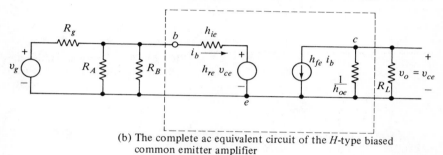

(b) The complete ac equivalent circuit of the *H*-type biased common emitter amplifier

FIGURE 4-8. Obtaining an equivalent circuit for an *H*-type common emitter amplifier.

4.5 Voltage Gain, Input and Output Impedances of a Current Biased Amplifier

A first look at the equivalent circuit in Fig. 4-8b is probably rather frightening to the newcomer, since it contains quite a few elements. At this stage it even seems like a more impossible task to solve a circuit with more than one transistor. However, all is not lost if certain procedures are followed. The first is to simplify the circuit to its most important properties, these being input and output impedances and voltage gain.

The student may also realize that the amplifiers studied thus far are basically of the same type and take on the name of the *common emitter amplifier* (implied in Chapter 3), since the emitter terminal is a common ac connection to both the input and output signals. There are two other basic types, and a knowledge of the input and output impedance and voltage gain allows us to understand more quickly the individual functions of these other configurations in a large circuit.

Let us now consider the circuit in Fig. 4-9 for the purpose of obtaining these properties. We shall assume that the capacitors C_i and C_o are sufficiently large to be short circuits to ac for the frequencies of interest. The equivalent small-signal ac circuit is shown in Fig. 4-10, and typical values for the elements are listed in Table 4-1.

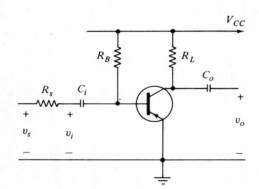

FIGURE 4-9 Common emitter amplifier driven with a nonideal source.

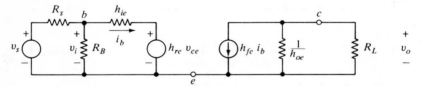

FIGURE 4-10 Equivalent circuit for a common emitter amplifier driven by a nonideal source.

TABLE 4-1 TYPICAL VALUES OF CIRCUIT ELEMENTS IN FIG. 4-10

Element	Typical Value
R_B	1.0 MΩ
h_{ie}	2.6 kΩ
h_{re}	2×10^{-4}
h_{fe}	100
$\dfrac{1}{h_{oe}}$	50 kΩ
R_L	2.2 kΩ

A value for R_s is not given, since this depends only on the source to be used. In some cases, such as laboratory signal generators, the value may be only a few hundred ohms or less, but it can also be as large as a few megohms for a crystal microphone. Let us assume that R_s is comparable in value to h_{ie}.

The circuit in Fig. 4-10 may immediately be simplified by the following approximations. Since R_B is typically 1 megohm and much larger than any other resistor in the network, it can be neglected. That is, very little ac current flows from the base to the emitter (in Fig. 4-10) through R_B, since it is so large. The next approximation is to neglect $1/h_{oe}$, which is typically much larger than R_L and is located in parallel with R_L. The parallel combination of $1/h_{oe}$ and R_L is essentially R_L. The network, approximated in this fashion, is shown in Fig. 4-11a.

There is still another simplifying approximation that can be made to lead to the simple equivalent circuit shown in Fig. 4-11b. We can neglect the voltage $h_{re}v_{ce}$, since it is very small in magnitude in comparison to the voltage v_{be}. We know that the voltage gain A_v, which is really v_{ce}/v_{be}, is of the order of several hundred. That is,

$$\frac{v_{ce}}{v_{be}} = -200 \quad \text{for a typical situation.} \tag{4-11}$$

Hence
$$|v_{ce}| = 200|v_{be}| \tag{4-12}$$
and
$$|h_{re}v_{ce}| = 200|h_{re}v_{be}| \tag{4-13}$$

Using a typical value for h_{re}, taken from Table 4-1, we obtain from Eq. (4-13),
$$|h_{re}v_{ce}| = 200 \times 2 \times 10^{-4}|v_{be}|$$
$$= 4 \times 10^{-2}|v_{be}| \ll |v_{be}|$$

That is, in most cases we can write
$$|h_{re}v_{ce}| \ll |v_{be}| \tag{4-14}$$

From Eq. (4-14), we see that the voltage due to h_{re} in the input is small in

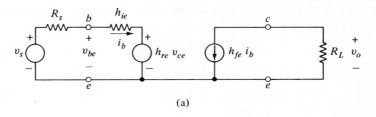

(a)

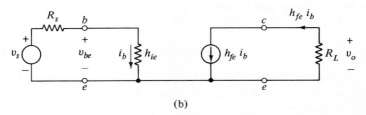

(b)

FIGURE 4-11 Simplification made upon the complete equivalent circuit (a) R_B and h_{oe} are neglected, (b) R_B, h_{oe}, and h_{re} are neglected.

comparison with voltage applied to the transistor and can be neglected. Notice that, in this example, the voltage due to source $h_{re}v_{ce}$ is about 4 percent of the applied voltage, v_{be}. Neglecting a voltage of this order of magnitude is a reasonable approximation to make in electronic design.

From Fig. 4-11b, we can now derive an expression for the voltage gain (A_v) of the amplifier, excluding the source resistance. That is,

$$A_v = \frac{v_{ce}}{v_{be}} = \frac{v_o}{v_{be}} = \frac{v_o}{v_i} \tag{4-15}$$

where

$$v_o = -h_{fe}i_b R_L \tag{4-16}$$

and, by KVL applied to the input loop,

$$v_i = v_{be} = h_{ie}i_b \tag{4-17}$$

so that dividing Eq. (4-16) by Eq. (4-17), we obtain

$$A_v = -\frac{h_{fe}R_L}{h_{ie}} = \frac{v_o}{v_i} \tag{4-18}$$

The expression for input impedance R_{in} can be obtained from Eq. (4-17). The input impedance is the ratio of the applied voltage at the input to the resulting input current, in this case i_b.

$$R_{in} = \frac{v_{be}}{i_b} = h_{ie} \tag{4-19}$$

To obtain the output impedance of this circuit, we need only find the Thévenin equivalent of the amplifier. The Thévenin voltage generator is simply the open-circuit voltage and is given by Eq. (4-16). By substituting the value for i_b in (4-16), from Eq. (4-17), we obtain the following expression for the open-circuit voltage v_{oc}:

$$\begin{aligned} v_{oc} = v_\theta = v_o &= -\frac{h_{fe}R_L}{h_{ie}}v_i \\ &= A_v v_{be} = A_v v_i \end{aligned} \tag{4-20}$$

The gain we have obtained has the ratio v_o/v_{be}, which is the output voltage over the base-to-emitter voltage.

To obtain the complete gain of the circuit, we include the effect of R_s by voltage division. The voltage v_{be} becomes

$$v_{be} = \frac{R_{in}}{R_s + R_{in}}v_s \tag{4-21a}$$

This procedure is clear and straightforward at a close examination of the input loop in Fig. 4-11b. Equation (4-21a) can be written in the form of a gain equation as follows:

$$\frac{v_{be}}{v_s} = \frac{R_{in}}{R_s + R_{in}} = \frac{h_{ie}}{R_s + h_{ie}} \tag{4-21b}$$

Recalling the gain equation, Eq. (4-18), of the transistor amplifier as

$$\frac{v_o}{v_{be}} = -\frac{h_{fe}R_L}{h_{ie}} \tag{4-22}$$

the product of Eq. (4-21b) and Eq. (4-22) yields the overall gain, including the source resistance:

$$\frac{v_o}{v_s} = \frac{v_o}{v_{be}} \frac{v_{be}}{v_s} = -\frac{h_{fe}R_L}{h_{ie} + R_s} \tag{4-23}$$

The output impedance, or the Thévenin impedance, becomes the ratio of the open-circuit voltage to the short-circuit current. We notice that applying a short-circuit to the output terminals in Fig. 4-11b results in a current equal in value to current generator applied to the short circuit. This is the direct result of the definition of an ideal current source given in Section 1.1:

$$R_{out} = Z_\theta = \frac{v_\theta}{i_{sc}} = \frac{-h_{fe}R_L i_b}{-h_{fe}i_b}$$

and

$$R_{out} = R_L \tag{4-24}$$

The Thévenin equivalent circuit of the amplifier is shown in Fig. 4-12. The student should memorize this result as well as Eq. (4-18), since this will be very useful in later work.

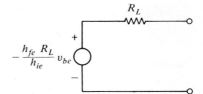

FIGURE 4-12 Thévenin equivalent of a common emitter amplifier stage.

Of course, with a little ingenuity, the solution for the circuit, as shown in Fig. 4-10, can be found without making any approximations. This will be done next to illustrate that very little accuracy is gained by such an approach. It should be understood that the results obtained are only to illustrate this point and that it is best to forget them once this task is accomplished. The procedure for the complete solution is as follows: Figure 4-10 is redrawn in Fig. 4-13 with v_s, R_s, and R_B replaced by a Thévenin equivalent consisting of v_g and R_g, where

$$v_g = \frac{R_B}{R_s + R_B} v_s \tag{4-25}$$

and

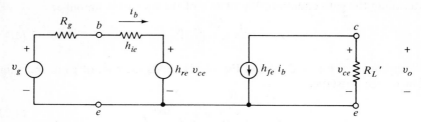

FIGURE 4-13 Circuit for the derivation of the exact expression for the voltage of a *CE* amplifier.

$$R_g = \frac{R_s R_B}{R_s + R_B} = R_s || R_B^* \tag{4-26}$$

The parallel combination of $1/h_{oe}$ and R_L is also replaced by a single resistor R_L', where

$$R_L' = \frac{R_L \frac{1}{h_{oe}}}{R_L + \frac{1}{h_{oe}}} = R_L || \frac{1}{h_{oe}} \tag{4-27}$$

As before, we can now obtain an expression for v_o:

$$v_o = -h_{fe} i_b R_L' = v_{ce} \tag{4-28}$$

At the input side we now write Kirchhoff's voltage law to obtain

$$v_g = (R_g + h_{ie}) i_b + h_{re} v_{ce} \tag{4-29}$$

Substituting Eqs. (4-25) and (4-28) into (4-29), we obtain

$$\frac{R_B}{R_s + R_B} v_s = (R_g + h_{ie}) i_b - h_{re} h_{fe} i_b R_L' \tag{4-30}$$

so that

$$v_s = \frac{R_s + R_B}{R_B} [(R_g + h_{ie}) - h_{re} h_{fe} R_L'] i_b \tag{4-31}$$

The expression for voltage gain becomes the ratio of Eq. (4-28) and (4-31):

$$A_v = \frac{-h_{fe} R_L'}{\frac{R_s + R_B}{R_B} [(R_g + h_{ie}) - h_{re} h_{fe} R_L']} \tag{4-32}$$

Two very apparent consequences immediately emerge from Eq. (4-32). The first is that it would be very difficult to remember an equation as large as this, and even more difficult for the designer to use. The second consequence is that by substituting in typical values, from Table 4-1, for R_s, R_B, h_{re}, h_{fe}, and R_L, we arrive at a solution that is very close to that obtained in Eq. (4-18).

*The symbol || will be used to denote parallel combinations of resistors. In this case, $R_s || R_B$ implies that R_s and R_B are in parallel.

Sec. 4.5 / Voltage Gain, Input and Output Impedances

To do this we shall look at the different groups of terms in Eq. (4-32). That is,

$$R'_L = \frac{R_L \frac{1}{h_{oe}}}{R_L + \frac{1}{h_{oe}}} = \frac{(2.2 \text{ k}\Omega)(50 \text{ k}\Omega)}{(52.2 \text{ k}\Omega)} = 2.11 \text{ k}\Omega$$

$$\frac{R_s + R_B}{R_B} = \frac{R_s + 10^6}{10^6} = 1 \quad \text{if } R_s < 10^4 \, \Omega$$

$$h_{re} h_{fe} R'_L = 2 \times 10^{-4} \times 10^2 \times 2.1 \times 10^3$$
$$= 4.2 \times 10 = 42 \, \Omega$$

$$R_g + h_{ie} = R_s || R_B + h_{ie} \simeq R_s + 2.6 \text{ k}\Omega$$

From these calculations it is apparent that R'_L is approximately equal to R_L. This, of course, depends on the "flatness" of the output characteristics of the transistor. The flatter the curves, the larger $1/h_{oe}$, and the better the approximation of neglecting $1/h_{oe}$. If the source resistance is small in comparison to R_B, then the term $(R_s + R_B)/R_B$ is certainly very close to unity, as indicated in the calculations. Finally, the term $h_{re} h_{fe} R'_L$ is certainly less than $R_s + h_{ie}$, since it is even less than h_{ie}. This approximation leads to a 2 per cent error at most, as indicated by the typical values.

In summary, the largest error that occurs in the voltage gain formula is the omission of $1/h_{oe}$ in parallel with R_L. However, Eq. (4-18) can be modified simply to

$$A_v = -\frac{h_{fe} R'_L}{R_s + h_{ie}} \tag{4-33}$$

with the R'_L to be understood as the parallel combination of R_L and $1/h_{oe}$ if necessary, and as simply R_L if $1/h_{oe}$ is large in comparison to R_L.

A similar procedure can now be applied to the calculation of input and output impedance, but it will be sufficient to say that the approximations used previously have been demonstrated to be quite good, and are useful within the limits of typical laboratory components and equipment. The approximate results are summarized in Table 4-2. These should be memorized.

TABLE 4-2 Symbols and Approximate Formulas for A_v, R_{in}, and R_{out}

Quantity	Symbol	Approximate Formula
Voltage gain	A_v^*	$-\dfrac{h_{fe} R'_L}{h_{ie}}$
Input impedance	R_{in}	h_{ie}
Output impedance	R_{out}	R_L

*Notice that the gain A_v is the ratio of the voltages v_{ce} and v_{be} and does not include the effect of source resistance. The effect of source resistance may be included, as was done in Eqs. (4-21) to (4-23).

4.6 Load Line Analysis

Another useful method in the analysis of transistor amplifier circuits is the method of *load line analysis*. This is a general technique for solving a simple loop consisting of numbers of linear resistors and a single nonlinear element whose voltage–current characteristics are well known. This method makes the technique ideal to solve for the *Q*-point and other properties of the transistor amplifier by graphical means.

Consider the amplifier stage shown in Fig. 4-14. Let us assume that v_i is first zero, or consider only the dc currents in the network. This gives rise to the following equation for the output loop:

$$V_{CC} = I_C(R_L + R_E) + V_{CE} \qquad (4\text{-}34a)$$

For the case in which v_i is zero, it can also be written as

$$V_{CC} = i_C(R_L + R_E) + v_{CE} \qquad (4\text{-}34b)$$

where it is understood that variational portions of i_C and v_{CE}, that is, i_c and v_{ce}, are zero. This equation is the locus of all the possible operating, or *Q*-points for the transistor circuit. Equation (4-34b) can now be written as

$$i_C = -\frac{1}{R_L + R_E} v_{CE} + \frac{V_{CC}}{R_L + R_E} \qquad (4\text{-}35)$$

We recall the equation of a straight line on the cartesian plane (see Fig. 4-15) to be

$$y = mx + b$$

where m is the slope of the line and b is the y-intercept. It is easy to see that Eq. (4-35) can be sketched in the v_{CE}, i_C plane, as shown in Fig. 4-16. The v_{CE}, i_C plane, of course, is the static output characteristic of the transistor.

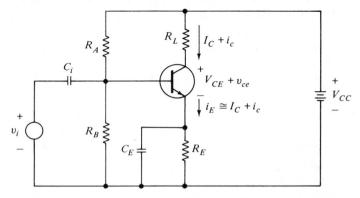

FIGURE 4-14   *CE* amplifier to be used for load line analysis.

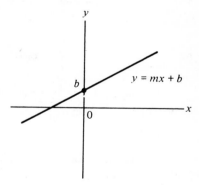

FIGURE 4-15 Equation of straight line plotted on the cartesian plane.

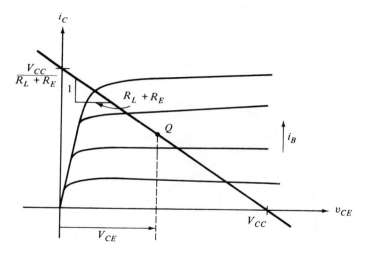

FIGURE 4-16 The dc load line plotted for the H-type biased CE amplifier.

It is easy to see that the v_{CE} intercept is V_{CC}, since the i_C intercept is $V_{CC}/(R_L + R_E)$ and the slope of the line is $-[1/(R_L + R_E)]$. The Q-point in Fig. 4-16 can be determined simply by calculating V_{CE}:

$$V_{CE} = V_{CC} - I_C(R_L + R_E) \tag{4-36}$$

For transistor amplifiers without an R_E, such as the one shown in Fig. 4-17, the load line analysis is the same except that the slope is $-(1/R_L)$ and the i_C intercept is V_{CC}/R_L as shown in Fig. 4-18.

The Q-point can now be determined either from a knowledge of I_B or I_C. The former is probably more likely, since $I_B \simeq V_{CC}/R_B$.

70 Chap. 4 / Biased Common Emitter Amplifier

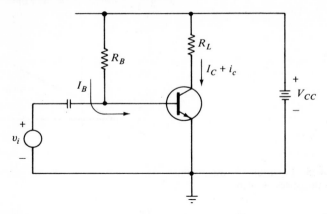

FIGURE 4-17 Constant-current biased *CE* amplifier to be used for load line analysis.

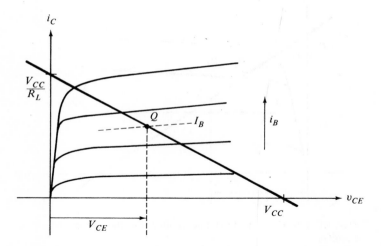

FIGURE 4-18 The dc load line and *Q*-point of a constant-current biased amplifier.

Some general rules can now be formulated for finding the load line and the *Q*-point, as follows:

1. Locate the value of V_{CC} along the v_{CE} axis.
2. Calculate the i_C intercept as the ratio of V_{CC} to the total resistance in the collector (or output) circuit.
3. Join these two points by a straight line.
4. Locate the *Q*-point. In the case of *H*-type biasing, this can be done by calculating V_{CE}, and, in the case of current source biasing, this

Sec. 4.6 / Load Line Analysis   71

is usually done by calculating I_B but can be done in the same manner as for H-type biasing.

The next step is to consider the amplifier under operation and to use the load line for ac analysis. The previous analysis applies to dc currents and voltages only and is referred to as *dc load line analysis*.

The *ac* load line is the locus of all instantaneous values of voltage and current during ac operation. The important thing to notice here is that the ac load (if one exists) must intersect the dc load line at Q-point, since setting the ac signal to zero leaves the transistor voltage and current at the Q-point. This means that the Q-point is the origin of the ac load line. Hence, for the output loop we can write for both Figs. 4-14 and 4-17 that the ac voltage drop v_{ce} is equal to the voltage drop across R_L, or

$$v_{ce} = -i_c R_L \tag{4-37}$$

This is true for Fig. 4-14, since C_E is assumed to be a large capacitor. Equation (4-37) can be rewritten as

$$i_c = -\frac{1}{R_L} v_{ce} \tag{4-38}$$

The equation is plotted in Fig. 4-19 for an *H*-type biased circuit, like the one shown in Fig. 4-14. Notice that the slope $(-1/R_L)$ of the ac load line is somewhat steeper than the slope of the dc load line. This is because the resistance at the emitter is shorted out by the capacitor C_E which reduces the ac resistance compared to the dc, as seen by the collector circuit.

Since the ac load is the locus of values of v_{ce} and i_c with Q-point as the origin, the values of v_{CE} and i_C can be determined for a given time-varying

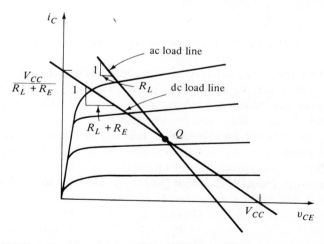

FIGURE 4-19   The ac and dc load line of an *H*-type biased amplifier.

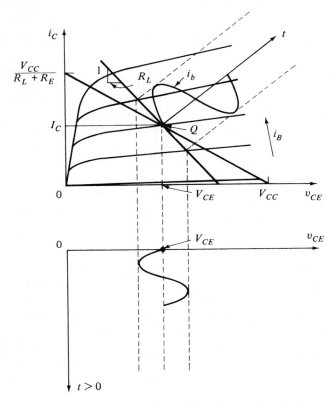

FIGURE 4-20 Use of the load line for signal analysis.

signal i_b. This procedure is easily understood by considering Fig. 4-20. In this figure, a time-varying signal i_b is applied. This will cause i_C and v_{CE} to vary in the manner indicated. Such an analysis is useful, since the maximum peak-to-peak output voltage can be determined.

If the Q-point is not suitably chosen, clipping of the output voltage and current can occur for a specified amount of i_b, as shown in Fig. 4-21. In this case, the Q-point was chosen too far up the dc load line. The base current i_b would have to be reduced, or a new Q-point would have to be chosen. This means that a new biasing circuit has to be designed. A good rule of thumb is to remember to choose the Q-point, as measured by V_{CE}, to be slightly under one half of V_{CC}. A value of $V_{CE} \simeq V_{CC}/2.1$ would allow the variations of v_{CE} to increase and decrease almost an equal amount.

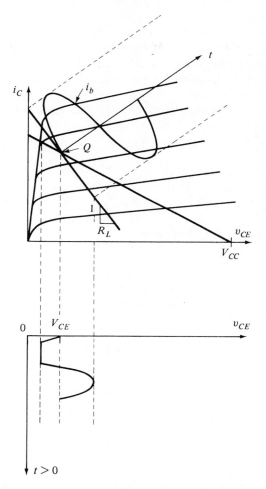

FIGURE 4-21 Load line analysis used to determine the amount of clipping in the amplified signal.

4.7 *H*-type Biased Amplifier

In Section 4.3, we discussed the voltage gain and the input and output impedances of a current biased amplifier, using equivalent circuits. It was found that many simplifying approximations could be made to the ac equivalent circuit without introducing too large an error. In fact, in the model for the transistor, the two significant parameters are h_{ie} and h_{fe}. The other two parameters, h_{oe} and h_{re}, are usually neglected. In the discussion that follows, we will do the same for this alternatively biased scheme.

Consider the amplifier shown in Fig. 4-22. Notice that the battery symbol for V_{CC} has been omitted. This is standard practice, and the V_{CC} designates the voltage between the designated point and the common, or ground, point. The simplified small-signal model for a transistor is shown in Fig. 4-23, with only the pertinent parameters. Now, using Fig. 4-23 and the fact that large capacitors and batteries are considered to be short circuits to ac, we can arrive at the simplified equivalent circuit shown in Fig. 4-24.

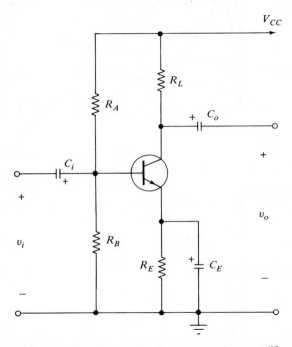

FIGURE 4-22 The H-type biased common emitter amplifier.

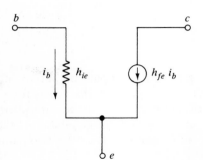

FIGURE 4-23 The simplified small signal model for a transistor.

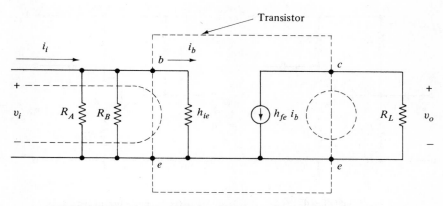

FIGURE 4-24 The simplified equivalent circuit for the common emitter amplifier.

We can immediately write an equation for the output voltage v_o:

$$v_o = -h_{fe} i_b R_L \qquad (4\text{-}39)$$

At the input loop, as designated by the dotted line, we can write a statement of KVL as

$$v_i = h_{ie} i_b \qquad (4\text{-}40)$$

Dividing Eq. (4-39) by Eq. (4-40), we obtain the expression for the voltage gain A_v:

$$A_v = \frac{v_o}{v_i} = -\frac{h_{fe} R_L}{h_{ie}} \qquad (4\text{-}41)$$

This result is the same as for the constant current biased circuit. This is to be expected, since the ac equivalent circuit is practically the same. The output impedance is the same as before:

$$R_{out} = R_L \qquad (4\text{-}42)$$

The only difference between the H-type biased circuit and the current biased circuit is in the input impedance. It is apparent that the input impedance is no longer h_{ie}, since R_A and R_B cannot be neglected. Hence the input impedance becomes the parallel combination of R_A, R_B, and h_{ie}.

$$R_{in} = \frac{v_i}{i_i} = R_A || R_B || h_{ie} \qquad (4\text{-}43)$$

4.8 Unbypassed Emitter Resistor Amplifier

The next amplifier circuit to be discussed is shown in Figs. 4-25a and 4-25b. It resembles the common emitter amplifier but has no bypass capacitor across

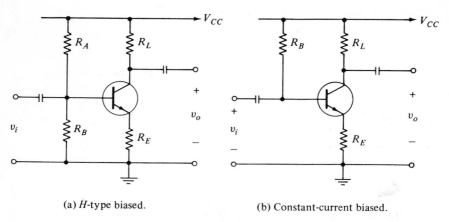

(a) *H*-type biased. (b) Constant-current biased.

FIGURE 4-25 The unbypassed emitter resistor amplifier.

R_E, the emitter resistor—hence the name. This amplifier can be biased by either the *H*-type biasing scheme or the constant current method. A study of this configuration is useful for two main reasons. First, a knowledge of the properties of this amplifier will allow us to detect a missing or faulty (open-circuited) capacitor. Second, this configuration may have properties that are useful and that could be used for meeting a design specification.

The approximate equivalent circuit for the amplifiers in Fig. 4-25 is given in Fig. 4-26. Notice that for the current biased amplifier, R_A is made infinite and the same equivalent circuit can be used. Notice that the current through the emitter resistor R_E is the sum of the two currents i_b and $h_{fe}i_b$.

The following equation can now be written for the output voltage:

$$v_o = -h_{fe}R_L i_b \qquad (4\text{-}44)$$

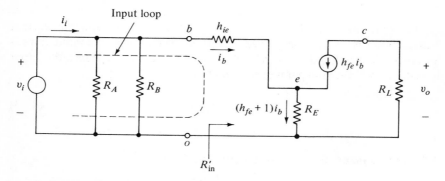

FIGURE 4-26 The simplified circuit for the unbypassed emitter resistor amplifier.

Sec. 4.8 / Unbypassed Emitter Resistor Amplifier   77

At the input we can write with the aid of KVL

$$v_i = i_b h_{ie} + (h_{fe} + 1)i_b R_E \qquad (4\text{-}45)$$

Dividing Eq. (4-44) by Eq. (4-45), we obtain the expression for voltage gain (A_v):

$$A_v = -\frac{h_{fe} R_L}{h_{ie} + (h_{fe} + 1)R_E} \qquad (4\text{-}46)$$

In most examples, h_{ie} is small in comparison to $(h_{fe} + 1)R_E$, so that Eq. (4-46) becomes

$$A_v = -\frac{h_{fe} R_L}{(h_{fe} + 1)R_E} \simeq -\frac{R_L}{R_E} \qquad (4\text{-}47)$$

The output impedance is the ratio of the open-circuit voltage to the short-circuit current at the output terminals:

$$R_{out} = \frac{\text{open-circuit voltage}}{\text{short-circuit current}}$$

$$= \frac{-h_{fe} i_b R_L}{-h_{fe} i_b} \qquad (4\text{-}48)$$

$$= R_L$$

The input impedance can be found by first finding the input impedance to the right of the terminals b and o in Fig. 4-26, and then adding R_A and R_B in parallel with this result. As shown in Fig. 4-26, the impedance to the right of the terminals is designated as R'_{in}. R'_{in} can be found from Eq. (4-45):

$$v_i = i_b h_{ie} + (h_{fe} + 1)i_b R_E \qquad (4\text{-}45)$$

$$R'_{in} = \frac{+v_i}{i_b} = h_{ie} + (h_{fe} + 1)R_E \qquad (4\text{-}49)$$

The input impedance to the amplifier is the ratio of input voltage v_i to input current i_i:

$$R_{in} = \frac{v_i}{i_i} = R_A || R_B || [(h_{fe} + 1)R_E + h_{ie}] \qquad (4\text{-}50)$$

If h_{ie} is much less than $(h_{fe} + 1)R_E$, then Eq. (4-50) becomes

$$R_{in} \simeq R_A || R_B || (h_{fe} + 1)R_E \qquad (4\text{-}51)$$

In summarizing, we list the results and make some observations that are apparent from these equations:

$$A_v \simeq -\frac{R_L}{R_E} \qquad (4\text{-}47)$$

$$R_{out} = R_L \qquad (4\text{-}48)$$

$$R_{in} = R_A || R_B || (h_{fe} + 1)R_E \qquad (4\text{-}51)$$

First of all, we notice that the voltage gain is the ratio of R_L to R_E, which would tend to give a lower gain than in the bypassed case, since, in *typically*

H-type biased amplifiers, R_E is only about one half to one twentieth as small as R_L. This makes the detection of a missing bypass capacitor easy, since gain will be between 2 and 20. In a current biased amplifier, R_E can take on any value; therefore, a full range of voltage gains—between 1 and several hundred—can be obtained by a proper selection of R_E and the use of Eq. (4-46).

Second, the output impedance is still R_L, as it has been for all the amplifiers we have studied thus far.

Finally, we notice that input impedance depends very much on the choice of R_A and R_B in the biasing. The term $(h_{fe} + 1)R_E$ is usually neglected in the calculation of input impedance, since it is much larger than R_A and R_B.

These points become clearer in the following examples:

Example 4-3: Given the amplifier circuit shown in Fig. 4-27, T_1 is a silicon *pnp* transistor, with $h_{FE} \simeq h_{fe} = 50$, and v_s is a sinusoidal signal whose frequency is 1 kHz. Calculate
(a) The voltage V_{CE}.
(b) The input impedance into the amplifier to the right of terminals b and o.
(c) The overall voltage gain v_o/v_s.

Solution:

(a) The first step is to assume that the voltage drop across R_E is negligible with respect to voltage across R_L and V_{CE}. Then it follows that $V_{BO} \simeq V_{BE} = 0.7$ V. Hence we can calculate base current I_B and I_C:

$$I_B \simeq \frac{V_{RB}}{R_B} = \frac{8.3 \text{ V}}{200 \text{ k}\Omega} = -41 \ \mu\text{A}$$

$$I_C = h_{FE}I_B = -50 \times 41 \mu\text{A} \simeq -2.1 \text{ mA}$$

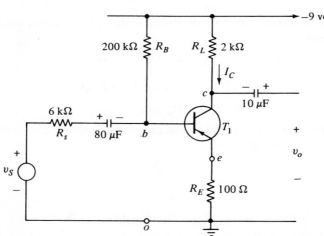

FIGURE 4-27 Figure for Ex. 4-3.

The voltage V_{CE} can now be calculated:

$$V_{CE} = V_{CC} - I_C(R_L + R_E)$$
$$= -9 \text{ V} + (2.1 \text{ mA})(2.1 \text{ k}\Omega)$$
$$= -4.6 \text{ V}$$

(b) The input impedance to the amplifier is given by

$$R_{in} \simeq R_B || (h_{ie} + h_{fe}R_E)$$

where

$$h_{ie} = \frac{0.026}{I_B} = \frac{0.026 \text{ V}}{42 \text{ }\mu\text{A}} = 620 \text{ }\Omega$$

and

$$h_{fe}R_E = 50 \times 100 \text{ }\Omega = 5 \text{ k}\Omega$$

Since R_B is much larger than $h_{ie} + h_{fe}R_E$, the input impedance R_{in} is simply

$$R_{in} \simeq h_{ie} + h_{fe}R_E$$
$$= 0.62 \text{ k}\Omega + 5 \text{ k}\Omega = 5.6 \text{ k}\Omega$$

(c) The overall voltage gain can be calculated by first finding the voltage gain v_o/v_{bo}.

$$A_v = \frac{v_o}{v_{bo}} = -\frac{h_{fe}R_L}{h_{ie} + h_{fe}R_E}$$
$$= -\frac{50(2 \text{ k}\Omega)}{5.6 \text{ k}\Omega} = -18$$

The next step is to calculate the gain v_{bo}/v_s by voltage division so that

$$\frac{v_o}{v_s} = \frac{v_{bo}}{v_s} \frac{v_o}{v_{bo}} = (0.48)(-18) = -8.6$$

It is interesting to make an additional observation here. Even though R_E is quite small, the calculation of the gain v_o/v_{bo} can be approximated by the formula

$$A_v = \frac{-R_L}{R_E} = -\frac{2 \text{ k}\Omega}{100 \text{ }\Omega} = -20$$

The resulting answer is within 10 percent of the more accurate calculation done in part (c). This is certainly within the experimental error acceptable in electronics.

Example 4-4: Given a 10 V dc power supply and a high-gain *npn* silicon transistor:
(a) Design an amplifier with a voltage gain of -10. The amplifier is to be driven by a voltage source whose output impedance is less than 10 Ω and whose frequency is 1 kHz.
(b) Calculate the input impedance of the resulting amplifier.

Solution:

(a) First of all, we notice that the voltage-gain requirement is quite low. This means that we would choose an unbypassed emitter resistor amplifier, since the voltage gain is given by

$$A_v = -\frac{R_L}{R_E}$$

Hence we choose $R_L = 10\ \text{k}\Omega$ and $R_E = 1\ \text{k}\Omega$, and proceed to bias the amplifier. The resulting circuit is shown in Fig. 4-28. All the pertinent dc voltages and currents are indicated on the circuit diagram.

(b) The input impedance of the amplifier is

$$R_{in} = R_A || R_B || (h_{fe} + 1)R_E$$
$$\simeq R_A || R_B$$

since for a high h_{fe} transistor, $(h_{fe} + 1)R_E$ is much larger than $R_A || R_B$.

$$R_{in} = \frac{(33.6\ \text{k}\Omega)(6.4\ \text{k}\Omega)}{33.6\ \text{k}\Omega + 6.4\ \text{k}\Omega} = 5.4\ \text{k}\Omega$$

Notice that the input signal voltage has been drawn as an ideal voltage source, since the 10-Ω source resistance is certainly negligible when compared to an input impedance of 5.4 kΩ.

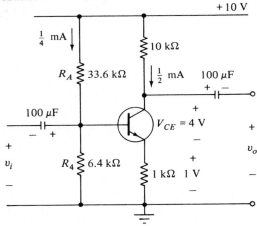

FIGURE 4-28 Figure for Ex. 4-4.

PROBLEMS

4.1 Bias a *pnp* silicon transistor with $h_{FE} = 50$ by using constant current biasing and a 12-V supply voltage.

4.2 What would the biasing voltage V_{CE} be if the *pnp* transistor in Problem 4.1

were replaced with a *pnp* germanium transistor with $h_{FE} \simeq 75$? Is the transistor still biased?

4.3 For a constant current biased transistor, such as the one shown in Fig. 4-1, what range of values can R_B be for the transistor to still be biased? The load resistor R_L is set at 2 kΩ, the supply voltage V_{CC} at 9 V, and the transistor has an h_{FE} of approximately 100.

4.4 Use the *H*-type biasing circuit to bias *pnp* silicon transistors with $40 < h_{FE} < 150$, using a 20-V supply.

4.5 If R_L is 5 kΩ and V_{CC} is 20 V, what values of I_E can be expected for an *H*-type biased transistor amplifier?

4.6 Use current-feedback biasing to bias an *npn* silicon transistor with $h_{FE} = 100$ from a 30-V supply.

4.7 A silicon *npn* transistor with $h_{FE} = 100$ is biased from a 10-V supply. The load resistor R_L is set at 2.5 kΩ and V_{CE} is 5 V.
  (a) What is the value of R_B needed to bias the transistor, using current-feedback biasing?
  (b) What will be the value of V_{CE} if the transistor in (a) is replaced by one whose h_{FE} is equal to 65? (*Hint:* Successive approximations will be necessary to solve this problem.)
  (c) If constant current biasing were used rather than feedback biasing, what would be the difference in V_{CE} upon replacing a transistor with $h_{FE} = 100$ by one that has $h_{FE} = 65$?

4.8 (a) Bias an *npn* transistor with $h_{FE} = 150$ and with $V_{CC} = 20$ V, using the circuits given in Fig. 4-6.
  (b) Compare the effects on V_{CE} of a change in h_{FE} to a value of 100 in these circuits with that change that would occur if constant current biasing were used.

4.9 (a) Obtain the equivalent circuit for the circuits in Problem 4.8, operating in the common emitter mode.
  (b) Calculate the input impedance, using the type of approximations discussed in Section 4.3.

4.10 (a) Obtain the complete equivalent circuit for a current-feedback biased amplifier in the common emitter configuration.
  (b) Obtain the approximate equivalent circuit by neglecting the effect of h_{re} and h_{oe} but not the effect of the biasing resistor R_B.
  (c) Derive an expression for voltage gain and input impedance for the circuit obtained in part (b).

4.11 A constant current biased common emitter amplifier is operated from 10-V supply. The load resistor R_L has a value of 4 kΩ and h_{FE} of the transistor is 80. What is the voltage gain that can be obtained from this amplifier if it is biased so that V_{CE} is approximately one half V_{CC}?

4.12 A signal generator with an open-circuit voltage of 15 mV peak to peak and an output impedance of 1,000 Ω is connected to the amplifier in Problem 4.11. What will be the resulting voltage at the output of the amplifier?

4.13 Derive the exact expression (using all the *h*-parameters) for the input impedance of constant current biased common emitter amplifier. How does this expression differ from Eq. (4-19)? What is a typical error if Eq. (4-19) is used rather than the exact expression?

4.14 (a) Sketch typical output characteristics for an *npn* transistor. The parameters h_{fe} and $1/h_{oe}$ are 50 and 40 kΩ, respectively, at the Q-point, which is situated at $V_{CE} = 10$ V and $I_E = 1$ mA. Assume as well that h_{FE} is 50 at the Q-point.
(b) Draw the load line for the amplifier shown in Fig. P4-14 on the sketch of part (a).
(c) Sketch the output voltage v_o if the peak-to-peak base current i_b is 10 μA. What value of v_i will be needed to produce such a current? (*Hint:* Calculate h_{ie} from $h_{ie} = v_T/I_B$.)

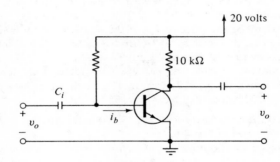

FIGURE P4-14

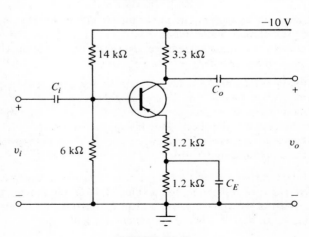

FIGURE P4-15

4.15 (a) Obtain the dc and the ac load lines for the amplifier shown in Fig. P4-15. Assume C_i, C_E, and C_o to be large capacitors.
  (b) What is the value of V_{CE} and I_C as obtained from the load line?
  (c) Calculate the input impedance for the amplifier shown in Fig. P4-15.
  (d) Calculate the voltage gain of the amplifier in Fig. P4-15 by use of formulas derived from equivalent circuits. Check this result by assuming a small voltage for v_i, calculating i_b, and obtaining the output voltage from the load line. (*Hint:* Sketch typical output characteristics, assuming a reasonable value of h_{fe} and h_{FE}.)

4.16 (a) Calculate the voltage gain and the input impedance for the amplifier shown in Fig. P4-16 for R equal to 0, 20, 40, and 60 Ω. Plot the results on a graph as a function of R. Assume that $h_{fe} = h_{FE} = 100$.
  (b) From the results obtained in part (a), what practical significance would the choice of R have in using a circuit such as this for design purposes?

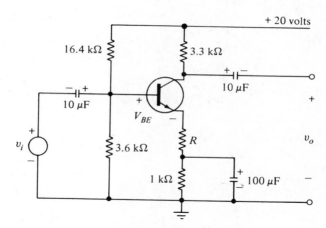

Frequency of $v_i = 1$ kHz

FIGURE P4-16

4.17 Design an amplifier with a voltage gain of -10. Specify all the components needed and the characteristics of the transistor needed for the design.

4.18 (a) Design an amplifier with a voltage gain of at least -200 and an output impedance of less than 5 kΩ. The amplifier must be capable of operating over a large range of temperatures which cause the h_{FE} of the available transistors to vary from 50 to 120. Assume h_{FE} to be approximately equal to h_{fe}.
  (b) Will the voltage gain vary over the extreme range of temperature? If so, how much?

4.19 Design an amplifier with a voltage gain of -50 and output impedance of 2 kΩ.

REFERENCES

[1] COMER, D. J., *Introduction to Semiconductor Circuit Design* (Addison-Wesley Publishing Co, Inc., Reading, Mass., 1968).

[2] CORNING, J. J., *Transistor Circuit Analysis and Design* (Prentice-Hall, Inc., Englewood Cliffs, N.J., 1965).

[3] JOYCE, M. V., and CLARKE, K. K., *Transistor Circuit Analysis* (Addison-Wesley Publishing Co., Inc., Reading, Mass., 1961).

[4] ANGELO, J. E., JR., *Electronics; BJTs, FETs, and Microcircuits* (McGraw-Hill Book Co., New York, 1969).

5

GENERIC EQUIVALENT CIRCUIT

In the vast field of electronics, one frequently meets two types of engineers: the *device engineer* and the *circuit engineer*. The device engineer studies the basic mechanisms of the transistor, and the circuit engineer studies the circuits that make the device operate in a desired fashion. The generic equivalent circuit, developed by device engineers, is based on solid-state physics, and some knowledge of this field helps the circuit engineer. The discussion that follows will give a brief, qualitative discussion of the semiconductor diode and the transistor from this point of view. It is by no means complete or rigorous, and the student is encouraged to do additional reading on his own to fill out the details.

5.1 Semiconductor Diode

A semiconductor can be defined as a substance whose electrical conductivity is between that of a conductor and an insulator, and also increases with temperature increase. Silicon is generally used at present, although germanium may be used and has been used to a large extent in the past. These two elements have covalent bonding between the atoms, and, in pure form, are found in a face-centered, diamond-like crystal structure. Like carbon, these elements belong to group IV in the periodic table. A one-dimensional illustration is shown in Fig. 5-1 for silicon. This illustration shows the outermost electrons forming a covalent bond by sharing these electrons to complete the outermost shell of eight electrons.

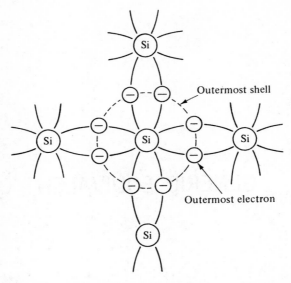

FIGURE 5-1 One-dimensional schematic of the crystal structure of silicon.

In pure germanium or silicon, conduction is possible by electrons leaving the outermost or valence shell (thanks to thermal energy) and moving to a higher energy level, called the *conduction band*. Electrons in this band are free to move if in the presence of an electric field. The vacancy left behind by such an electron is called a *hole*. If an electric field is applied, as shown in Fig. 5-2, the hole appears to move in the direction of the applied field, and the electron, in the opposite direction. This type of conduction in a semiconductor material is called *intrinsic conduction*.

To improve the conduction in an intrinsic semiconductor, such as germanium or silicon, impurities (dopants) from group III or group V elements can be added to the material. Group III elements have three electrons in the

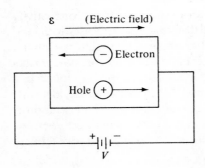

FIGURE 5-2 Direction of motion of a hole and an electron due to an electric field.

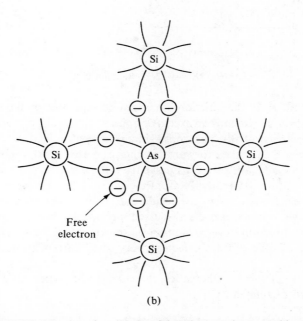

FIGURE 5-3 Atomic structure of doped germanium and silicon.

outermost orbit, and result in an extrinsic semiconduction as shown in Fig. 5-3a. Indium is usually used. Notice that, since there are only three electrons in the outermost orbit, additional holes are created in the valence band. An intrinsic semiconductor material in which these additional holes have been created by a trivalent dopant is called a *p-type material*. Similarly, creation of *n-type material* is done by adding arsenic, a group V type of material, in which free electrons are created. The resulting crystal structure is shown in Fig. 5-3b. The process of adding these impurities to the silicon or germanium is called *doping*.

Conduction in doped semiconductor material occurs under two mechanisms. *Drift* is the motion of holes or free electrons due to the influence of an electric field, and *diffusion* is the motion of holes or free electrons from a region of high concentration to a region of low concentration.

To construct a semiconductor diode we fuse a *p*-type material to an *n*-type material, as shown in Fig. 5-4a. The symbol for the diode is shown in Fig. 5-4b. This type of diode is called a *junction diode*, and the region that separates the *p*-type material from the *n*-type material is called the *junction*.

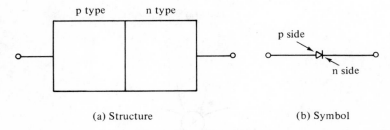

(a) Structure   (b) Symbol

FIGURE 5-4 A *pn* junction diode.

The operation of the junction diode can be explained by the fact that the *p*-side of the diode contains an excess of holes owing to doping, and the *n*-side contains an excess of free electrons. Under an electric field, in the direction shown in Fig. 5-5a (called forward biasing), the excess holes move from the *p*-side to the *n*-side, and electrons move from the *n*-side to the *p*-side, causing electric current to flow easily. Notice that in Fig. 5-5a we have indicated that a few free electrons exist in the *p*-side and a few holes exist in the *n*-side. These are the intrinsic carriers discussed previously. Now, under a field in the opposite direction (*reverse biasing*), as shown in Fig. 5-5b, only the intrinsic carriers are able to flow, causing a very small current to flow across the junction.

By the use of some mathematics, the voltage and current in the diode can be shown to be related by

$$i_D = I_R(e^{(q/mkT)v_D} - 1) \tag{5-1}$$

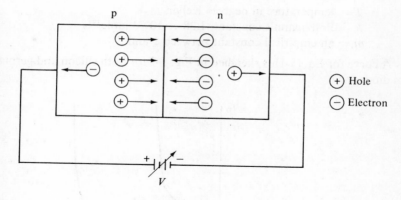

(a) Forward-biased diode.

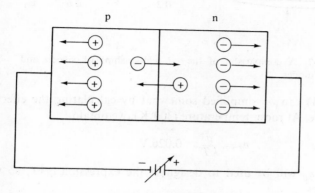

(b) Reverse-biased diode.

FIGURE 5-5 Biasing the *pn* junction diode.

where

i_D = current through the diode with reference direction shown in Fig. 5-6

v_D = voltage across the diode with reference direction as shown in Fig. 5-6

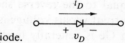

FIGURE 5-6 Voltages and current reference directions of a diode.

I_R = reverse saturation current (or leakage current)
q = electron charge 1.6×10^{-19} coulomb

T = temperature in degrees Kelvin
k = Boltzmann's constant, 1.38×10^{-23} joule/°K
m = an empirical constant between 1 and 2

A curve for Eq. (5-1) is sketched in Fig. 5-7 for both silicon and germanium diodes.

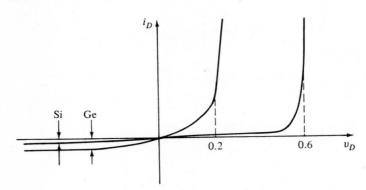

FIGURE 5-7 A comparison of the v-i relationships of silicon and germanium diodes.

Equation (5-1) can be simplified somewhat by calculating the effect at room temperature. At room temperature (300°K), we obtain

$$v_T = \frac{q}{kT} \simeq 0.026 \text{ V}$$

The symbol v_T will be used to designate the expression q/kT, so that Eq. (5-1) becomes

$$i_D = I_R(e^{v_D/mv_T} - 1) \tag{5-2}$$

It also is interesting to note that when v_D is a negative quantity, we can write an expression for the current i_D to be

$$i_D = I_R(e^{-|v_D|/m(0.026)} - 1)$$

so that

$$i_D \simeq -I_R \tag{5-3}$$

provided $|v_D| > m(0.026)$.

When $|v_D|$ becomes much larger than $m(0.026)$, then the current becomes equal to the reverse saturation (or leakage) current. A plot of the reverse current versus voltage is sketched in Fig. 5-8. Notice that the leakage current in Ge is generally much larger than in Si. Although not drawn to scale, it would be well to note that, *in practice* the reverse current in silicon diodes is in the order of nanoamperes (10^{-9} A) and microamperes (10^{-6} A) for germanium diodes.

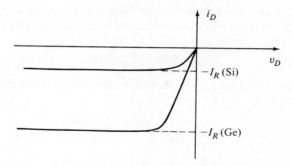

FIGURE 5-8 Reverse characteristics of a silicon and germanium diode.

When v_D is positive Eq. (5-2) can be written approximated as

$$i_D \simeq I_R e^{v_D/mv_T}$$

provided

$$v_D > m(0.026)$$

This approximation is very good where v_D is greater than about 0.1 V. It is also interesting to note [although not apparent from Eq. (5-2)] that the upward break for silicon diodes is about 0.6 V, and 0.2 V for germanium diodes.

It is also useful, at this time, to calculate the slope of the v_D–i_D curve in the forward-biased condition. This is an important result, since it gives the incremental (small-signal) resistance of the diode in the forward direction r_d by

$$i_D \simeq I_R e^{v_D/mv_T} \tag{5-4}$$

then

$$\frac{\partial i_D}{\partial v_D} = \text{slope of characteristic curve}$$

$$= \frac{I_R}{mv_T} e^{v_D/mv_T}$$

$$\simeq \frac{1}{mv_T} I_D = \frac{1}{r_d}$$

where

$$r_d = \text{incremental resistance of the diode}$$

$$= \frac{mv_T}{I_D}$$

$$\simeq \frac{m(0.026)}{I_D} \text{ at room temperature}$$

Hence the slope is the reciprocal of incremental resistance and proportional to the average current through the diode I_D (see Fig. 5-9).

$$\frac{\Delta i_D}{\Delta v_D} = \frac{1}{r_d} = \frac{I_D}{m(0.026)} \tag{5-5}$$

Notice that the larger I_D, the smaller r_d, and hence the steeper the slope of the characteristic.

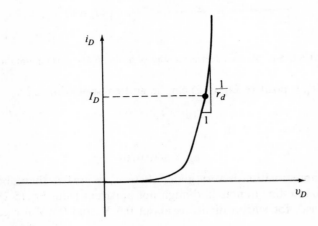

FIGURE 5-9 Forward characteristics of a diode illustrating the incremental resistance r_d.

5.2 Use of Semiconductor Diode

At this stage, our primary reason for studying the semiconductor is to lay a foundation for further study of the transistor. It is apparent that there is a strong resemblance between the v_D-i_D curves and the input characteristics of a transistor. It is this aspect of the diode that we shall dwell upon later.

In this section, however, we shall discuss one of the primary uses of diodes —that is, for making rectifiers. This will also be discussed in greater detail in a later chapter.

The first step is to approximate the diode characteristic in Fig. 5-10a by the *ideal diode* characteristic shown in Fig. 5-10b. The interpretation of this curve is simple. If a positive voltage is applied across the diode, the diode acts like a short circuit. This is represented by the vertical line along the positive i_D axis, for $v_D = 0$. If a negative voltage is applied across the diode, no current flows, and the diode acts as an open circuit. This is represented by a horizontal line on the negative v_D axis, for $i_D = 0$.

To make a rectifier now, one need only connect the diode to a sinusoidal

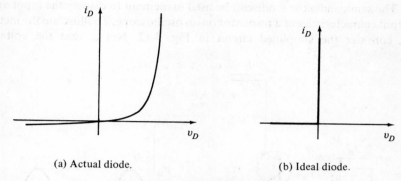

(a) Actual diode.   (b) Ideal diode.

FIGURE 5-10 Diode characterization.

source v_1 and a load resistor R_L, as shown in Fig. 5-11a. The resulting voltage across R_L, $v_{RL}(t)$ is plotted in Fig. 5-11b. The reason for the resulting wave form is obvious, since the diode D_1 conducts (the diode is a short circuit) when v_1 is positive, and D_1 does not conduct when v_1 is negative.

If a physical semiconductor diode is used rather than the ideal diode in the circuit, the voltage $v_{RL}(t)$ would be the wave form sketched with the dotted line in Fig. 5-11b, and would not be sinusoidal. For many applications, however, the ideal diode approximation is sufficient for design.

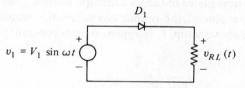

(a) Rectifier circuit using a semiconductor diode

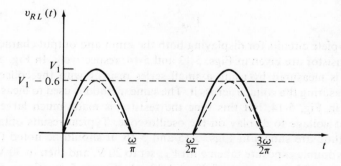

(b) Output wave form of the rectifier circuit

FIGURE 5-11 Rectification by a *pn* diode.

The semiconductor diode can be used in a circuit to display the input and output characteristics of a transistor on an oscilloscope. To illustrate the method, consider the simplified circuit in Fig. 5-12. Notice that the voltage

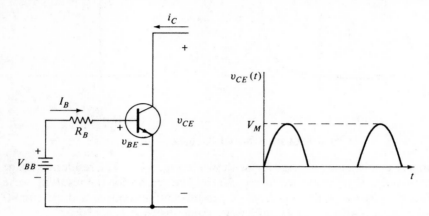

FIGURE 5-12 Simplified circuit for obtaining a display of output characteristics.

v_{CE} varies between zero and V_M. For a given value of I_B, i_C will increase and decrease according to the v_{CE}–i_C relationship for the given transistor. If i_C were displayed by the vertical plates of the oscilloscope while v_{CE} was applied to the horizontal plates, a plot of the output characteristics would be displayed. Notice that in the above setup, I_B is approximately constant and given by

$$I_B \simeq \frac{V_{BB}}{R_B} \qquad (5\text{-}6)$$

provided

$$V_{BB} > V_{BE}$$

Complete circuits for displaying both the input and output characteristics of a transistor are given in Figs. 5-13 and 5-14, respectively. In Fig. 5-13, the current is measured by using a small series resistor with the collector and then measuring the voltage across it. The same method is used to measure base current in Fig. 5-14, but this time the resistor is made much larger to get sufficient voltage to display on the oscilloscope. Typical results obtained by this method are shown in Figs. 5-15 and 5-16. It should be noted that Fig. 5-15 is a double exposure taken with V_{BB} set to 20 V, and then to 30 V, giving two values of base current of 20 and 30 μA. Figure 5-16 is also a double exposure with V_{CE} set at two values, 5 and 15 V. The change in the input characteristic is virtually undetectable.

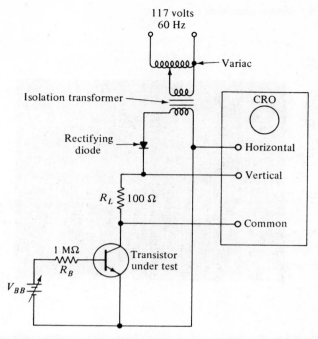

FIGURE 5-13 Complete circuit for displaying output curves for an *npn* transistor.

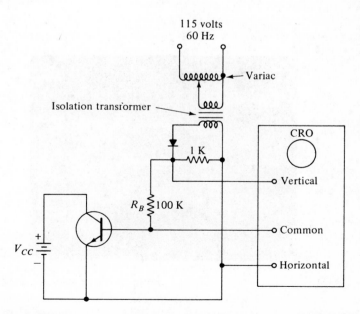

FIGURE 5-14 Complete circuit for displaying the input characteristics of a transistor.

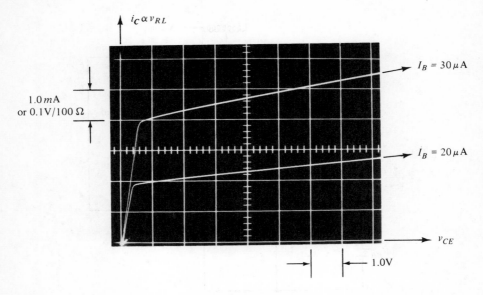

FIGURE 5-15 Photograph of output characteristics obtained by curve tracer of Figure 5-13.

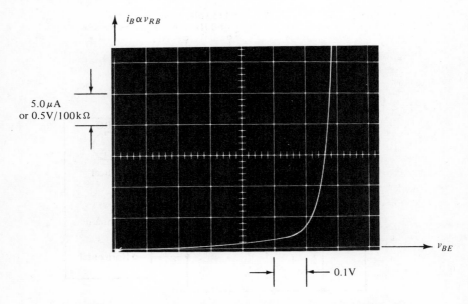

FIGURE 5-16 Photograph of input characteristics obtained by curve tracer of Figure 5-14.

5.3 Extension of Diode Equation to the Transistor

We have already learned how to bias a transistor and use it as a common emitter amplifier. Let us now take a second look at the voltages in a properly biased transistor. Figure 5-17 illustrates a biased silicon *pnp* transistor, with

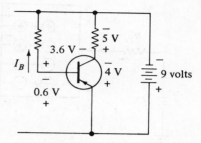

FIGURE 5-17 dc voltages in a biased *pnp* transistor.

dc voltages as indicated. Notice that for a biased *pnp* transistor the base is made more negative than the emitter and the collector is more negative than the base, since

$$V_{CB} = V_{CE} - V_{BE} \tag{5-7}$$

A physical transistor is made of three pieces of semiconductor material with *n*-type material sandwiched between two pieces of *p*-type material, as shown in Fig. 5-18a. The emitter, base, and collector leads are connected to each portion. The base region is made very thin. Notice that the transistor has two junctions—the emitter base junction and the base collector junction. An illustration of an *npn* transistor is shown in Fig. 5-18b.

Using the previous observations as to the polarity of proper voltages for

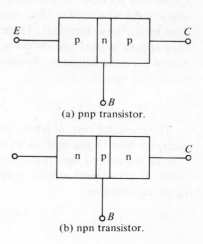

FIGURE 5-18 Simplified pictorial representation of the junction transistor.

biasing, the transistor can be biased as shown in Fig. 5-19. Notice that V_{EE} supplies a positive voltage between emitter and base (the base is more negative than the emitter), and V_{CC} supplies a negative voltage between collector and base.

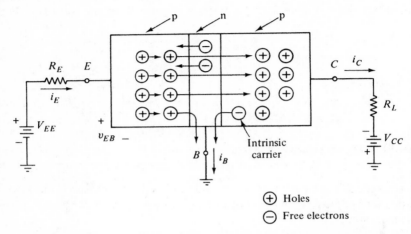

FIGURE 5-19 A biased *pnp* transistor in the common base configuration showing the internal flow of holes and electrons.

Since the positive terminal of V_{EE} is connected to the emitter, or *p*-type material, the emitter base junction is forward biased. It acts like a forward biased diode, and voltage between emitter and base is

$$V_{EB} \simeq 0.6 \text{ V} \quad \text{(for silicon)} \tag{5-8}$$

Under the presence of this bias, the holes in the emitter (*p*-type material) move to the base region. Since the base region is very narrow, few of the holes flow to the base terminal. These holes are allowed to diffuse through the base region toward the collector, where they are attracted by the negative potential on the collector. About 1 percent of the holes from the emitter constitute the base current:

$$i_B \simeq 0.01 i_E \tag{5-9}$$

If the emitter current is set equal to zero, some intrinsic electrons from the collector flow to the base as well. Notice that the base collector junction is reverse biased and that this current will be the same as the leakage current for a reverse-biased silicon diode. The two effects can be combined in a single equation as

$$i_C = \alpha i_E + I_{CBO} \tag{5-10}$$

where

i_C = the total collector current

i_E = the total emitter current

I_{CBO} = leakage current between collector and base when the emitter is open-circuited

α = proportionality constant that defines the portion of carriers that reach the collector from the emitter region

It is fairly easy to see that quantity α will be about 0.99 (very close to unity), since only about 1 percent of carriers from the emitter comprises the base current.

The explanation of the operation of an *npn* transistor is identical to a *pnp* transistor, except that the batteries are reversed and the emitter current, which flows, is comprised of electrons moving from the emitter to the base. Hence all the currents are reversed (using conventional current as a reference).

The action of the *pnp* transistor can also be represented by an equivalent circuit. This circuit is shown in Fig. 5-20. In this figure, D_1 represents the

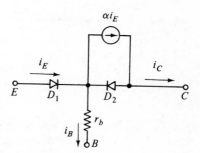

FIGURE 5-20 A diode equivalent circuit for a *pnp* transistor.

forward-biased emitter base junction, and D_2 represents the reverse-biased base collector junction. The dependent current generator αi_E is the portion of emitter current that goes to the collector. The resistor r_b has been added as an ohmic base resistance, which makes up the lead resistance and the resistance of very narrow base region. It is due primarily to the fact that the base region is very narrow. The value of r_b is usually between 5 to 40 Ω for a silicon transistor; 20 Ω is usually taken for a typical value. In a germanium transistor, this value can be as high as 200 Ω. A good typical value for a germanium transistor is 50 Ω.

The equivalent circuit shown in Fig. 5-20 is useful as an equivalent circuit for the transistor in the common base configuration. It is *not* a small-signal model but, rather, represents the total effect of the transistor. This is the reason why the symbols for currents are written in lowercase with uppercase subscripts. I_{CBO}, the collector base leakage current, is constant and therefore is written in uppercase letters with uppercase subscripts.

5.4 Generic and Tee Models for a Transistor

In Section 5.3, a model representing the total current was obtained for a transistor. It is now quite easy to construct an equivalent small-signal model. This model will be called the generic model, since it is derived from physical considerations. The equivalent circuit obtained previously is shown in Fig. 5-21, with the currents separated into the dc and ac components. Notice that an additional symbol is used to denote the internal base region in the transistor. This node has been labeled X.

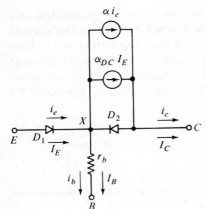

FIGURE 5-21 The diode equivalent circuit for a transistor illustrating both the ac and dc currents in the device.

To obtain the small-signal equivalent circuit, we will have to look at all the voltage–current relationships in each branch on an incremental basis:

1. The forward-biased diode on an incremental basis can be replaced by the incremental resistance of the forward-biased diode. This resistance is called r_e. The value of r_e will depend upon the current through it, I_E. Hence

$$r_e = \frac{mv_T}{I_E} \tag{5-11}$$

2. The resistor r_b remains the same, since it acts the same to ac as to dc.
3. Diode D_2 must be replaced by the incremental resistance of a reverse-biased diode. This, of course, is the reciprocal of the slope of the diode curve in the third quadrant, as shown in Fig. 5-22. We shall call this resistance r_c (the incremental collector resistance). The value of r_c is very large. In practice, this value ranges from about 1 to 10 MΩ.
4. Finally, the dependent current generator $\alpha_{DC}I_E$ is neglected, since

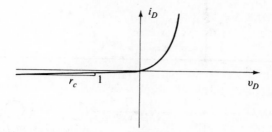

FIGURE 5-22 The incremental resistance of a reverse biased diode.

it is constant, and the current generator αi_e is left. Notice that the α and α_{DC} need not be the same value. This takes into account changes of α with bias conditions.

The final small-signal ac equivalent circuit is shown in Fig. 5-23. This equivalent finds its primary use in an amplifier in the common base configuration but can be used for the common emitter configuration, if desired. The reason that it is more useful for the common base configuration is because the current generation αi_e is dependent on the emitter current, which is the input current in the common base configuration.

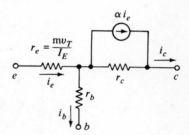

FIGURE 5-23 Common base generic small signal model for a transistor.

To make the generic equivalent circuit more useful for the common emitter configuration, it can be changed to have the dependent generator written as a function of i_b. This will be done next in several simple steps.

1. Thévenin's theorem is used to replace the combination of αi_e and r_c. The resulting circuit is shown in Fig. 5-24a.
2. Since $i_e = i_b + i_c$, the voltage generator $\alpha r_c i_e$ can be written as

$$\alpha r_c i_e = \alpha r_c (i_b + i_c)$$
$$= \alpha r_c i_b + \alpha r_c i_c \quad (5\text{-}12)$$

This means that this single generator can now be replaced by two equivalent generators, as shown in Fig. 5-24b.
3. Since the voltage generator $\alpha r_c i_c$ has the current i_c passing through it but with the tail of the current arrow corresponding to a nega-

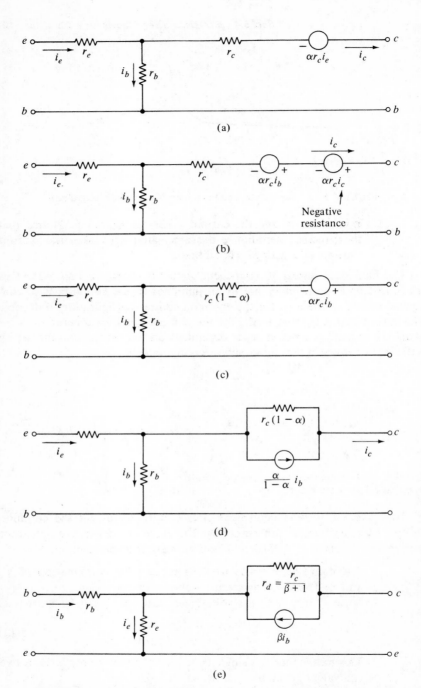

FIGURE 5-24 Series of circuits illustrating the method of obtaining the common emitter generic equivalent from the common base circuit.

tive voltage, we recognize this as a negative resistor and lump it with r_c, as shown in Fig. 5-24c.

4. The final step is to convert the voltage generator $\alpha r_c i_b$ and resistor $r_c(1 - \alpha)$ to a Norton equivalent. The current generator becomes

$$\frac{\alpha r_c i_b}{r_c(1 - \alpha)} = \frac{\alpha}{1 - \alpha} i_b \qquad (5\text{-}13)$$

as shown in Fig. 5-24d.

5. The term $\alpha/(1 - \alpha)$ is usually written as β. Hence

$$\beta = \frac{\alpha}{1 - \alpha} \qquad (5\text{-}14)$$

The resistor $r_c(1 - \alpha)$ can also be written in terms of β from Eq. (5-14).

$$\beta(1 - \alpha) = \alpha$$
$$\beta = \alpha(1 + \beta)$$

and

$$\alpha = \frac{\beta}{\beta + 1} \qquad (5\text{-}15)$$

Now,

$$(1 - \alpha) = 1 - \frac{\beta}{\beta + 1} = \frac{1}{\beta + 1} \qquad (5\text{-}16)$$

Hence the resistor $r_c(1 - \alpha)$ is usually written as

$$r_d = \frac{r_c}{\beta + 1} \qquad (5\text{-}17)$$

The reason for this is that the term β is much easier to measure than α. This will be seen in later discussion.

The final step is to reverse all the current directions in the circuit and twist the base and emitter leads into alternate positions. The final circuit is shown in Fig. 5-24e. There are several observations that can be made about this final equivalent circuit. First of all, it has the same topology (elements are connected in the same fashion) as the equivalent circuit derived from z-parameters in Chapter 3. Therefore, it must obviously have the same elemental values. The equivalent circuit, as derived from z-parameters, is known as the *tee*, or *r-parameter model*. Second, the value of β must be quite large, since β is equal to $\alpha/(1 - \alpha)$ and α is a number very close to unity. Finally, this must be a low-frequency model, since all the impedances in the circuit are resistors. This was discussed previously in Chapter 3.

5.5 h- and r-Parameter Interrelationships

In Section 5.4 we have shown the method for arriving at the *common emitter equivalent tee model* from a physical point of view, and we have also shown that this is the same model that can be obtained from z-parameter two-port theory. Some of the parameters in the tee model are known very precisely, such as r_b and r_e.

$$r_b = \text{ohmic base resistance (typically 20 } \Omega\text{)} \tag{5-18}$$

$$r_e = \frac{mv_T}{I_E}$$

However, the parameters r_d and β are not as well known. We can summarize the results we have thus far as

$$\beta = \frac{\alpha}{1-\alpha} \quad \text{where } \alpha \text{ is very close to unity} \tag{5-19}$$

and

$$r_d = \frac{r_c}{\beta + 1} \quad \text{where } r_c \text{ is the reverse incremental resistance of a diode*} \tag{5-20}$$

At this point α is completely unknown, and small errors in estimating α produces a large error in β. r_d is very difficult to measure.

In the case of h-parameters, the output characteristics are easy to obtain either from plotting or from the manufacturer, and the h_{fe} and h_{oe} can be found quite accurately from the curves. The terms h_{ie} and h_{re} are much more difficult to obtain.

It appears that a relationship between the h- and r-parameters may help to eliminate this difficulty. In fact, for the most part, it does. Consider the equivalent tee model, shown in Fig. 5-25. Using KVL, we can write two equations for Fig. 5-25:

$$v_{be} = (r_b + r_e)i_b + i_c r_e \tag{5-21}$$

$$v_{ce} = i_c(r_d + r_e) + i_b r_e - \beta i_b r_d \tag{5-22}$$

Recalling the definition of h_{oe} and using Eq. (5-22), we can write

$$h_{oe} = \frac{i_c}{v_{ce}}\bigg|_{i_b=0} = \frac{1}{r_d + r_e} \tag{5-23}$$

Similarly, using the definition for h_{re} and Eq. (5-21) and (5-22) we obtain

$$h_{re} = \frac{v_{be}}{v_{ce}}\bigg|_{i_b=0} = \frac{r_e}{r_d + r_e} \tag{5-24}$$

*Note that this r_d is not same as the r_d introduced in Section 5.1, an unfortunate choice of symbols.

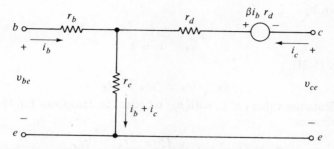

FIGURE 5-25 Equivalent tee model for the transistor in the common emitter configuration.

Some approximations are in order now. Since $1/h_{oe}$ is the reciprocal of the slope of the output characteristics and is of the order of 50 kΩ, and, since r_e is of the order of 25 Ω or so, we can write that

$$h_{re} \cong \frac{r_e}{r_d} \tag{5-25}$$

as well as

$$h_{oe} \cong \frac{1}{r_d} \tag{5-26}$$

From Eqs. (5-25) and (5-26), we now have a better knowledge of the values of some of the parameters that were difficult to measure before. For example, $1/h_{oe}$ is easy to measure, and therefore we now know that r_d is approximately the same. Since we have an acceptable formula for r_e, and now know r_d, we have a better estimate of h_{re}.

Let us now calculate h_{fe} and h_{ie} in a similar fashion. h_{fe} is given by

$$h_{fe} = \frac{i_c}{i_b}\bigg|_{v_{ce}=0} \tag{5-27}$$

Setting v_{ce} equal to zero in Eq. (5-22,) we obtain

$$0 = i_c(r_d + r_e) + i_b r_e - \beta i_b r_d \tag{5-28}$$

or

$$i_c(r_d + r_e) = i_b(\beta r_d - r_e) \tag{5-29}$$

Hence

$$h_{fe} = \frac{i_c}{i_b}\bigg|_{v_{ce}=0} = \frac{\beta r_d - r_e}{r_d + r_e} \tag{5-30}$$

Using the same approximation as before, that $r_e \ll r_d$, we obtain

$$h_{fe} \cong \beta \tag{5-31}$$

h_{ie} is given by

$$h_{ie} = \frac{v_{be}}{i_b}\bigg|_{v_{ce}=0}$$

If, in Eq. (5-21),

$$v_{be} = (r_b + r_e)i_b + i_e r_e \tag{5-21}$$

we substitute the value for i_c, with v_{ce} set equal to zero from Eq. (5-29), Eq. (5-21) becomes

$$v_{be} = (r_b + r_e)i_b + \left(\frac{\beta r_d - r_e}{r_d + r_e}\bigg|_{v_{ce}=0}\right)i_b r_e \tag{5-32}$$

Solving for h_{ie} from Eq. (5-32), we obtain

$$h_{ie} = r_b + r_e\left(1 + \frac{\beta r_d - r_e}{r_d + r_e}\right)$$
$$= r_b + (h_{fe} + 1)r_e \quad \text{(exactly)}$$

or

$$h_{ie} \simeq r_b + (\beta + 1)r_e \simeq \beta r_e \tag{5-33}$$

As before, from Eqs. (5-31) and (5-33), we are able to obtain a more accurate knowledge of the parameters that previously were difficult to measure. Now β is approximately the same as h_{fe}. Also, h_{ie} may be obtained from a knowledge of r_e and β. For most cases, r_b is negligible and the best we usually do is guess at its value, using the typical value of 20 Ω for silicon.

For the alternative form of the tee model (common base configuration) shown in Fig. 5-26, we have two additional parameters, α and r_d. These are easy to obtain, since we already have equations for these in terms of β and r_c:

$$\alpha = \frac{\beta}{\beta + 1} \tag{5-34}$$

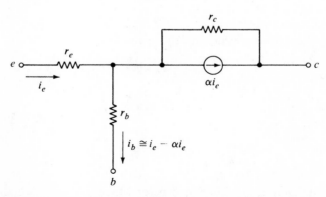

FIGURE 5-26 Tee model for a transistor in the common base configuration.

and
$$r_c = (\beta + 1)r_d \tag{5-35}$$
These two new parameters can be calculated from the β and r_d.

A summary of the results obtained in this section is shown in Table 5-1, an important table that should more or less be memorized. A quick observation of it shows that two parameters h_{fe} and h_{oe} must be obtained from measurement (usually from the output characteristics). This makes the setup shown in Fig. 5-13 quite important, since the output characteristics are quickly displayed. Commercial curve tracers such as the Tektronix 575 or 576 can be very useful but not absolutely necessary, since an estimated value of parameters is usually accurate enough for many design procedures.

TABLE 5-1 METHODS FOR OBTAINING COMMON TRANSISTOR PARAMETERS

Parameter	Easiest Solution
r_e	calculate from $r_e = \dfrac{mv_T}{I_E}$
h_{fe}	measure h_{fe} from the output characteristics
h_{oe}	measure h_{oe} from the output characteristics
β	assume $\beta \simeq h_{fe}$
r_b	estimate a value of 20 Ω (higher for Ge)
h_{ie}	calculate from $h_{ie} \simeq r_b + (\beta + 1)r_e$
r_d	assume $r_d \simeq \dfrac{1}{h_{oe}}$
h_{re}	calculate from $h_{re} \simeq \dfrac{r_e}{r_d}$
α	calculate from $\alpha = \dfrac{\beta}{\beta + 1}$
r_c	calculate from $r_c = (\beta + 1)r_d$

Example 5-1: h_{fe} and $1/h_{oe}$ have been measured for an *npn* transistor as 80 and 30 kΩ, respectively. The transistor is to be biased with $I_E = 2$ mA. Obtain the equivalent tee model for
(a) The common emitter configuration.
(b) The common base configuration.

Solution:

(a) First, r_b is estimated to be 20 Ω.
$$\beta \simeq h_{fe} = 80$$
and
$$r_d \simeq \frac{1}{h_{oe}} = 30 \text{ k}\Omega$$
$$r_e = \frac{mv_T}{I_E} \simeq \frac{0.026}{2 \times 10^{-3}} = 13 \text{ }\Omega \quad \text{(assuming that } m = 1\text{)}$$

The complete equivalent circuit is shown in Fig. 5-27.

108   Chap. 5 / Generic Equivalent Circuit

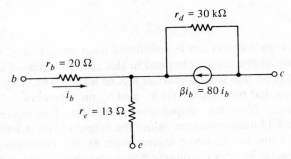

FIGURE 5-27   Circuit for Ex. 5-1.

(b) For the common base configuration, we shall need the values of r_c and α.

$$r_c = (\beta + 1)r_d \simeq 80(30 \text{ k}\Omega)$$
$$= 2.4 \text{ M}\Omega$$

$$\alpha = \frac{\beta}{\beta + 1} = \frac{80}{80 + 1} = 0.988$$

The complete equivalent circuit is shown in Fig. 5-28.

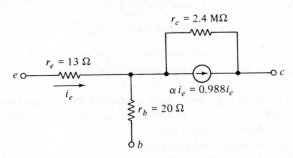

FIGURE 5-28   Completed equivalent circuit for Ex. 5-1.

Example 5-2: A *pnp* transistor has a value of $r_d = 35$ kΩ for a specified Q-point of $V_{CE} = 5$ V and $I_E = 2$ mA. Calculate the approximate value of h_{re}.

Solution:

The solution of this problem is interesting, since it will confirm a typical value of h_{re} given in Chapter 3. $r_e = 13$ Ω from Example 5-1.

$$h_{re} \simeq \frac{r_e}{r_d} = \frac{13}{35 \times 10^3} = 3.7 \times 10^{-4}$$

5.6 Properties of Common Emitter Amplifier Using Tee Model

We have already calculated the input impedance, output impedance, and voltage gain of a common emitter amplifier using h-parameters. We shall now use the r-parameter model to obtain the same results.

Consider the H-type biased amplifier shown in Fig. 5-29a. The complete equivalent circuit can be obtained, as before, by replacing all large capacitors and batteries with short circuits, and by replacing the transistor with its equiv-

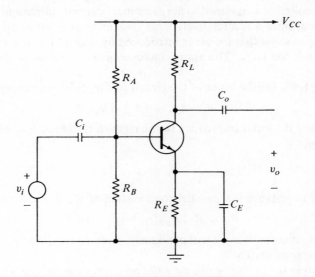

(a) Common emitter amplifier.

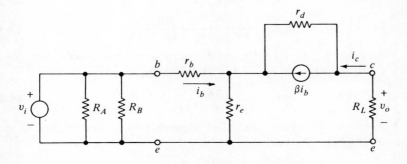

(b) Complete equivalent circuit using the equivalent Tee model.

FIGURE 5-29 CE H-type biased amplifier and equivalent circuit.

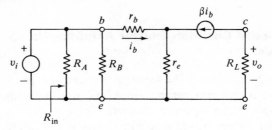

FIGURE 5-30 Simplified equivalent circuit using the Tee model

alent circuit. The resulting circuit is shown in Fig. 5-29b. As was previously done, the collector is assumed to act as an ideal current source and $r_d\,(=1/h_{oe})$ is neglected, since it is a relatively large resistance. It is left as an exercise for the reader to show that the error introduced by doing this for a typical set of values is not too large. The approximate equivalent circuit is shown in Fig. 5-30.

Using KVL for the input of the circuit in Fig. 5-30, we can write

$$v_i = i_b r_b + (\beta + 1)r_e i_b \tag{5-36}$$

To calculate the input impedance to the right of terminals b-e, we divide Eq. (5-36) by i_b:

$$\frac{v_i}{i_b} = r_b + (\beta + 1)r_e \tag{5-37}$$

The input impedance is a parallel combination of R_A, R_B, and v_i/i_b or

$$R_{in} = R_A \| R_B \| [r_b + (\beta + 1)r_e] \tag{5-38}$$

This is the same result as we had previously, since we recognize $r_b + (\beta + 1)r_e$ as being approximately h_{ie}.

To obtain the voltage gain, we write an expression for the output voltage and divide it by Eq. (5-36):

$$v_o = -\beta i_b R_L \tag{5-39}$$

$$A_v = \frac{v_o}{v_i} = -\frac{\beta R_L}{r_b + (\beta + 1)r_e} \tag{5-40}$$

Notice that Eq. (5-40) is the same result that we had using h-parameters, since $h_{fe} \simeq \beta$ and $h_{ie} \simeq r_b + (\beta + 1)r_e$. That is,

$$A_v \simeq -\frac{\beta R_L}{r_b + (\beta + 1)r_e} \simeq -\frac{h_{fe} R_L}{h_{ie}} \tag{5-41}$$

Equation (5-40) can be simplified somewhat by observing that, in most cases, $r_b \ll (\beta + 1)r_e$, so that

$$A_v \simeq -\frac{\beta R_L}{(\beta + 1)r_e} = -\alpha \frac{R_L}{r_e}$$
$$\simeq -\frac{R_L}{r_e} \quad \text{since } \alpha \simeq 1 \tag{5-42}$$

Equation (5-42) is in a very compact form and illustrates that the voltage gain of a common emitter amplifier is almost independent of the current gain β, depending only on R_L and the bias current I_E, since $r_e = mv_T/I_E$.

The output impedance can now be calculated very quickly:

$$R_{out} = \frac{v_{oc}}{i_{sc}} = \frac{-\beta i_b R_L}{-\beta i_b} = R_L \tag{5-43}$$

This is the same result we had when h-parameters were used.

Example 5-3:

Given: A constant current biased common emitter amplifier.

Required: Since the voltage gain of a common emitter amplifier depends only on R_L and the biasing current, what will be the relationship between the voltage gain A_v and voltage V_{CC}, which provides the bias?

Solution:

The magnitude of the gain of the amplifier shown in Fig. 5-31 is

$$|A_v| = \left|\frac{R_L}{r_e}\right| = \left|\frac{R_L I_E}{mv_T}\right| \tag{5-44}$$

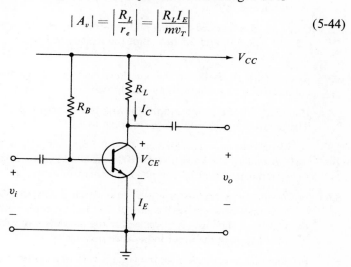

FIGURE 5-31 Circuit for Ex. 5-3.

If the amplifier is biased correctly, we can write, as a very approximate equation, that the dc voltage across R_L is one half the battery voltage V_{CC}.

$$|R_L I_C| \simeq \tfrac{1}{2}|V_{CC}|$$

or

$$|V_{CC}| = 2|R_L I_C|$$

since

$$I_E \simeq I_C$$

$$|R_L I_E| = \tfrac{1}{2}|V_{CC}| \tag{5-45}$$

Substituting Eq. (5-45) into Eq. (5-44) we obtain

$$|A_v| = \frac{\frac{1}{2}|V_{CC}|}{mv_T} \simeq 20\,|V_{CC}| \qquad \text{for } m = 1 \tag{5-46}$$

The result of Example 5-3 is interesting, since it shows that increasing the supply voltage V_{CC} has a tendency to increase the gain of the amplifier. This is not strictly true, since the bias point for the collector-to-emitter voltage does not always have to be selected at one half the supply voltage.

PROBLEMS

5.1 (a) Assuming I_R to be 5×10^{-8} A, plot the diode equation, Eq. (5-1), in the forward direction for four or five values of v_D up to about 0.6 V. Assume m to be equal to 1.0.
(b) Plot the same curve for five values of v_D in the negative direction up to about -5.0 V.

5.2 Calculate the forward incremental resistance r_d of a diode for I_D equal to 0.5, 1.0, 1.5, 2.0, and 2.5 mA. Plot the resistance r_d on a graph as function of I_d.

5.3 Devise a circuit similar to the one shown in Fig. 5-13 to display the forward characteristic of a diode on an oscilloscope.

5.4 Using Eq. (5-15), plot a graph showing the values of β as a function of α. The parameter α should range from 0.9 to 1.0.

5.5 A silicon transistor is biased at $I_E = 0.5$ mA and $V_{CE} = 5$ V. The parameters β and r_d are known to be 100 and 30 kΩ, respectively. Calculate all the parameters listed in Table 5-1 for this transistor.

5.6 (a) A silicon *npn* transistor is available with a β between 50 and 150. Design two amplifiers, one with a gain of 200 and the other with a gain of -200.
(b) Calculate the input impedance for the amplifiers of part (a). What range of values will the input impedance take on?

5.7 (a) Obtain an expression for the voltage gain of a common emitter amplifier, using *r*-parameters, without neglecting the effect of r_d.
(b) Using typical values for the *r*-parameters of a transistor, what is the typical error introduced by neglecting r_d?

5.8 Using the results of Example 5-3, give the maximum peak-to-peak input voltage v_i that can be used for a constant current biased amplifier before clipping occurs.

5.9 (a) Using the results of Example 5-3, give the supply voltage that would be necessary to obtain a gain of -60.
(b) Design an *H*-type biased amplifier with the supply voltage obtained in part (a). Calculate the gain. Explain why the gain is not exactly -60.

5.10 (a) Using r-parameters, design an amplifier with a gain of 150 and an input impedance of 2 kΩ. Specify the β of the transistor necessary to meet the specifications.

(b) Suppose that the transistor that was available for the circuit designed in part (a) had a β 50 percent higher than the specified value. What effect would the use of transistor in the circuit have on the desired results? How would you correct this result?

5.11 (a) In Fig. P5-11, both the diode D_1 and the transistor T_1 are silicon devices. R_E is made equal to R_B. What effect, if any, does the diode D_1 have on the ac operation of the amplifier circuit?

(b) Does the diode D_1 have any effect on the dc operation (biasing) of the circuit?

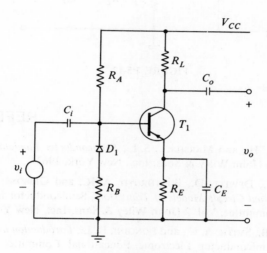

FIGURE P5-11

5.12 (a) Bias the amplifier shown in Fig. P5-12 with R_L equal to R_E.

(b) Using r-parameters, and neglecting the effect of r_d, calculate the gain v_{o_1}/v_i.

What is the gain v_{o_2}/v_i? Assume C_i, C_{o_1}, and C_{o_2} to be large capacitors. (*Hint:* R_E is usually much larger than r_e.)

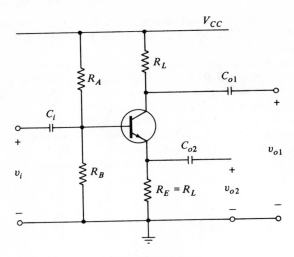

FIGURE P5-12

REFERENCES

[1] SEIDMAN, A. H., and MARSHALL, S. L., *Semiconductor Fundamentals, Devices and Circuits* (John Wiley & Sons, Inc., New York, 1963).

[2] GRAY, P. E., DEWITT, D., BOOTHROYD, A. R., and GIBBONS, J. F., *Physical Electronics and Circuit Models of Transistors*, Semiconductor Electronic Educational Committee, Vol. 2 (John Wiley & Sons, Inc., New York, 1964).

[3] ALDER, R. B., SMITH, A. C., and LONGINI, R. L., *Introduction to Semiconductor Physics*, Semiconductor Electronic Educational Committee, Vol. 1 (John Wiley & Sons, Inc., New York, 1964).

[4] LEVINE, S. N., *Principles of Solid-State Microelectronics* (Holt, Rinehart and Winston, Inc., New York, 1963).

[5] GIBBONS, J. F., *Semiconductor Electronics* (McGraw-Hill Book Co., New York, 1966).

6

COMMON BASE
AND COMMON COLLECTOR AMPLIFIERS

In previous chapters, we discussed the common emitter (CE) amplifier, which has, as the name implies, the emitter as a common connection to the input and output circuit. The base terminal is used for the input and the collector is used as the output. This circuit is shown symbolically in Fig. 6-1a. There are two other amplifier configurations that can be used. These are shown in Figs. 6-1b. and 6-1c, and are called the *common base* and the *common collector amplifiers*. (Other connections of the terminals, the source and the load, are possible, but these are not used.) The common emitter, common base, and common collector amplifiers are usually abbreviated as the CE, CB, and CC amplifiers.

6.1 Biasing the CB Amplifier

In Section 5.3 the equivalent tee model was derived on the basis of a CB amplifier. To bias any transistor for amplifier operation, the emitter base junction must be forward biased and the collector base junction be reverse biased. When this is done, the voltage–current properties of the transistor are those previously discussed and, therefore, the same equivalent circuits can be used.

The biasing schemes that were studied in Chapter 4 (Sections 4.1 to 4.3) may, therefore, be used to operate a transistor in the CB configuration. All we need to do is to apply our input to the emitter and measure our output at the collector, making sure that the base is common to both the input and

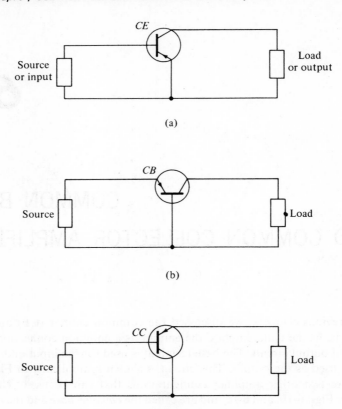

FIGURE 6-1 The three basic configurations: (a) the common emitter, (b) the common base, and (c) the common collector amplifier.

the output. This is done in Fig. 6-2 with the three basic biasing schemes that were studied in Chapter 4.

In each circuit in Fig. 6-2, C_B, a large capacitor, is placed between the base and ac ground. Remember that to ac a battery is a short circuit. The capacitor C_i couples the input to the emitter, and C_o is a coupling capacitor for the output.

6.2 Electronic Properties of the CB Amplifier

As in the case of the CE amplifier, we shall now calculate the input impedance, output impedance, and voltage gain of the CB amplifier. Consider the common base amplifier in Fig. 6-3a. The equivalent circuit for the amplifier is shown in Fig. 6-3b, and was obtained in the usual way. Notice that R_A

Sec. 6.2 / Electronic Properties of the CB Amplifier   117

(a) H-type biased CB amplifier

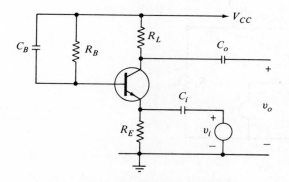

(b) Constant-current biased CB amplifier

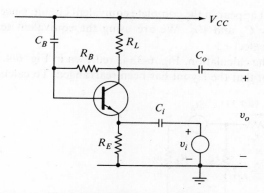

(c) Feed back biased CB amplifier

FIGURE 6-2 The biased CB amplifier.

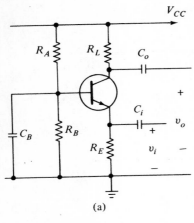

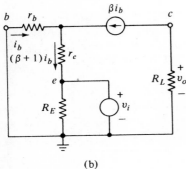

(a) The biased *CB* amplifier, and
(b) its simplified equivalent circuit.

FIGURE 6-3 CB amplifier and simplified equivalent circuit.

and R_B do not appear in the complete equivalent circuit, since they are shorted to ground by C_B and V_{CC}. We are using the equivalent tee model and, as before, we neglect r_d.

To ease the calculation, Fig. 6-3b is redrawn in Fig. 6-4. This is the same circuit, except that the layout has been rearranged. To calculate the gain, the

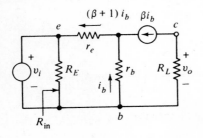

FIGURE 6-4 CB amplifier after simple rearrangement.

output voltage v_o is needed:

$$v_o = -\beta i_b R_L \tag{6-1}$$

At the input we can write, with the aid of KVL,

$$v_i = -[(\beta + 1)i_b r_e + i_b r_b] \tag{6-2}$$

Dividing Eq. (6-1) by Eq. (6-2), we obtain the expression for voltage gain:

$$A_v = \frac{\beta R_L}{(\beta + 1)r_e + r_b}$$
$$\simeq +\frac{R_L}{r_e} \tag{6-3}$$

since $\quad r_b \ll (\beta + 1)r_e$

Notice that the voltage gain for the CB amplifier is as large, in magnitude, as the CE emitter amplifier but is opposite in phase. The + sign indicates that input and output voltages are in phase. This is as expected, since the difference between the CB and CE amplifiers is the manner in which the input is applied. In the CB amplifier, applied voltage v_i is between the base and the emitter, except that the + terminal is applied to the emitter rather than to the base, as in the CE amplifier.

To calculate the input impedance of the CB amplifier, we first find the input impedance to the right of the terminals e–b in Fig. 6-4 and then add R_E in parallel with this result. The input impedance to the right of these terminals is the ratio of the applied voltage v_i and the input current $-(\beta + 1)i_b$. Using Eq. (6-2), we obtain

$$\frac{v_i}{-(\beta + 1)i_b} = r_e + \frac{r_b}{\beta + 1}$$
$$\simeq r_e \quad \text{since } r_b \ll r_e(\beta + 1) \tag{6-4}$$

The input impedance becomes

$$R_{in} \simeq r_e \parallel R_E \tag{6-5}$$

For most biasing circuits, R_E is much larger than r_e and Eq. (6-5) becomes

$$R_{in} \simeq r_e \tag{6-6}$$

The input impedance of a CB amplifier is rather low, since it is given by

$$r_e = \frac{mv_T}{I_E}$$

However, the CB amplifier can be useful for terminating coaxial cables, whose characteristic impedance is usually 52 or 75 Ω. By adjusting I_E to $\frac{1}{2}$ or $\frac{1}{3}$ mA, the desired 52 or 75 Ω can easily be obtained.

The output impedance of the CB amplifier is obtained by dividing the

120   Chap. 6 / Common Base and Common Collector Amplifiers

open-circuit output voltage by the short-circuit output current:

$$R_{out} = \frac{v_{oc}}{i_{sc}} = \frac{-\beta i_b R_L}{-\beta i_b} = R_L \qquad (6\text{-}7)$$

The resulting Thévenin equivalent circuit is shown in Fig. 6-5.

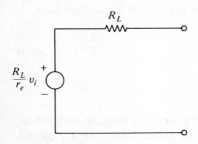

FIGURE 6-5 The Thévenin equivalent of a CB amplifier.

The electronic properties of the CB amplifier could have been obtained using the other two models that we have for the transistor, the CE h-parameter and CB equivalent tee model. The steps for using the CE h-parameter model will now be outlined, but the use of the CB equivalent tee model is left as an exercise.

1. The common base amplifier is represented by the approximate equivalent circuit, as shown in Fig. 6-6.
2. The voltage gain is now calculated:

$$v_o = -h_{fe} R_L i_b$$
$$v_i = -i_b h_{ie}$$
$$A_v = \frac{v_o}{v_i} = +\frac{h_{fe} R_L}{h_{ie}} \qquad (6\text{-}8)$$

This has the same magnitude as the CE amplifier and, upon substituting the values of h_{ie} and h_{fe} in terms of the r-parameters, we obtain

$$A_v = +\frac{R_L}{r_e} \qquad (6\text{-}9)$$

3. The input impedance is calculated next:

$$R_{in} = R_E \| \frac{h_{ie}}{h_{fe} + 1} \qquad (6\text{-}10)$$

and, substituting, as in step 2,

$$R_{in} = R_E \| \frac{r_b + (\beta + 1)r_e}{(\beta + 1)}$$
$$\simeq r_e \qquad (6\text{-}11)$$

Sec. 6.2 / Electronic Properties of the CB Amplifier

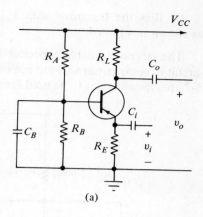

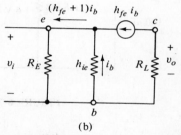

FIGURE 6-6 CB amplifier and h-parameter equivalent circuit.

(a) The CB amplifier.
(b) The ac equivalent circuit using the simplified h-parameter circuit.

4. The output impedance becomes

$$R_{out} = \frac{-h_{fe}R_L i_b}{-h_{fe}i_b} = R_L \tag{6-12}$$

Example 6-1: Given an *npn* silicon transistor with $h_{fe} = 60$, design an amplifier with an input impedance of 52 Ω and a voltage gain of 130. The amplifier is to operate from a 9-V supply.

Solution:

(a) To obtain an input impedance of 52 Ω, we use a CB amplifier with $I_E = 0.5$ mA, since

$$R_{in} \simeq r_e = \frac{0.026}{0.5 \times 10^{-3}} = 52 \text{ Ω}$$

(b) To obtain the voltage gain of 130, we solve for R_L from

$$A_v = \frac{R_L}{r_e}$$

$$R_L = A_v r_e = 130 \times 52 \simeq 6.8 \text{ kΩ}$$

(c) Bias the transistor with $V_{CC} = 9$, $R_L = 6.8$ kΩ, and $I_E = 0.5$ mA, as shown in Fig. 6-7.

The largest available capacitor is used for C_i, since it must be a short circuit in comparison to input impedance of r_e. The capacitors shown in Fig. 6-7 are typical values to be used for audio frequencies.

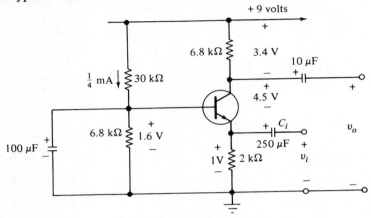

FIGURE 6-7 Figure for Ex. 6-1.

6.3 Two-Stage Amplifiers

The need for a two-stage amplifier may arise for various reasons. One of these may be that the gain of a one-stage amplifier may not be large enough. If this is the case, all that need be done is to connect a suitable amplifier stage to the output and further amplify the signal.

To illustrate the method for cascading two stages of amplification, consider, first of all, the common emitter amplifier, shown in Fig. 6-8. The ampli-

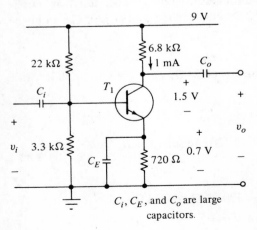

C_i, C_E, and C_o are large capacitors.

FIGURE 6-8 A biased CE amplifier.

fier has been biased at $V_{CE} = 1.5$ V and $I_C = 1$ mA. The gain of the amplifier is found to be approximately 250. Assume that the desired gain is about 500. The obvious answer is, of course, to add a second low-gain stage, an unbypassed emitter resistor amplifier. Such an amplifier stage is shown in Fig. 6-9.

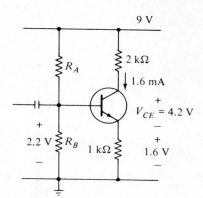

FIGURE 6-9 A biased transistor amplifier with an unbypassed emitter resistor.

In Fig. 6-9, resistors R_A and R_B have not been specified. Their function is to provide a 2.2-V drop between the base and the ground. They are really unnecessary, since such a voltage already exists from collector to ground, shown in Fig. 6-8. The collector dc current through T_1 is certainly much larger than the base current of T_2, and the use of the first stage to bias the second, as shown in Fig. 6-10, would certainly not change the bias point T_1. This method of connecting two stages is called *direct coupling*.

Another reason for more than one stage of amplification is the problem

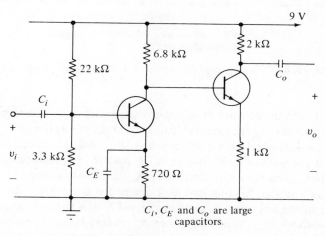

FIGURE 6-10 Direct coupling of two amplifier stages.

encountered when an external load is connected. Consider the amplifier and its equivalent circuit shown in Fig. 6-11. The voltage gain of this amplifier can be calculated as being

$$\frac{v_o}{v_i} = \frac{-R_L}{r_e} \quad \text{where } R_L = R'_L \| R_{ext}$$

$$= \frac{-1.65 \text{ k}\Omega}{26 \text{ }\Omega} = -63 \tag{6-13}$$

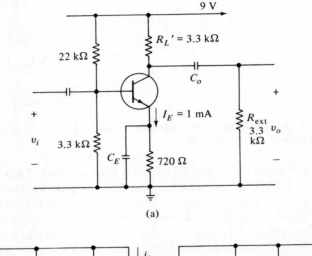

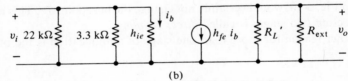

(a) A CE amplifier with an external load, and
(b) The equivalent circuit

FIGURE 6-11 A CE amplifier with external load with its equivalent circuit.

Notice that with the absence of R_{ext} the voltage gain doubles and goes up to about 130. Decreasing the external load resistor decreases the gain. In order that the gain does not decrease substantially, R_{ext} should be much larger than R'_L, so that the parallel combination of R'_L and R_{ext} is, essentially, R'_L. The effect of decreasing gain with decreasing R_{ext} is called *loading*.

Another way that loading may be explained is to consider the Thévenin equivalent of the amplifier, as given in Fig. 6-12. A_v is the voltage gain of the amplifier without R_{ext}. Notice that when R_{ext} is very large, the output voltage

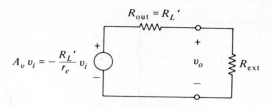

FIGURE 6-12 The Thévenin equivalent of a CE amplifier with an external load.

v_o is the voltage $A_v v_i$. As R_{ext} is decreased, the voltage across the output impedance R_L tends to decrease the voltage v_o. The voltage division theorem can be used to calculate the output voltages as follows:

$$v_o = \frac{R_{ext}}{R'_L + R_{ext}} A_v v_i \qquad (6\text{-}14)$$

The voltage gain becomes

$$\frac{v_o}{v_i} = \frac{R_{ext}}{R'_L + R_{ext}} \frac{R'_L}{r_e} = \frac{R'_L \| R_{ext}}{r_e} \qquad (6\text{-}15)$$

When loading does occur, more than one stage may be necessary to obtain the desired amount of amplification. However, neither the CB nor the CE amplifier are useful for having large external loads connected to their terminals. It is the common collector (CC) amplifier that is best suited for this purpose; it will be considered in Section 6.4.

Loading occurs not only when an external load is added to the circuit but must also be taken into account when a second stage of amplification is added. The question then arises: Why was the effect of loading not accounted for in the example in Fig. 6-10? The answer is that the second stage had a high-enough input impedance so that the loading effect of the second stage was negligible. If we recall the formula for the input impedance of an unbypassed resistor amplifier as

$$R_{in} = R_A \| R_B \| [(h_{fe} + 1)R_E] \qquad (6\text{-}16)$$

we see that in the absence of R_A and R_B in Fig. 6-10, Eq. (6-16) becomes

$$R_{in} = (h_{fe} + 1)R_E \qquad (6\text{-}17)$$

For typical values of h_{fe} and R_E equal to 1 kΩ, the input impedance becomes very large. Therefore, the second stage in the example does not present a load to the first stage.

Example 6-2: The two-stage amplifier shown in Fig. 6-13 is being driven from a signal source with an output impedance of 25 Ω. The transistors being used are *npn* silicon with a β of approximately 50. The capacitors C_i, C_B, C_E, and C_o are assumed to be large. Calculate the overall voltage gain v_o/v_s.

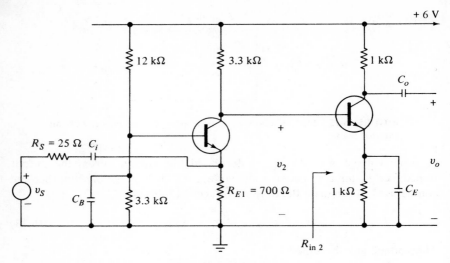

FIGURE 6-13 Figure for Ex. 6-2.

Solution:

(a) The first step in the solution of a problem of this type is to establish all the dc voltages and currents, and enter them into a circuit diagram, such as the one shown in Fig. 6-14. Notice that the capacitors are not shown in this diagram, since they have no effect on these *dc* quantities. The voltage

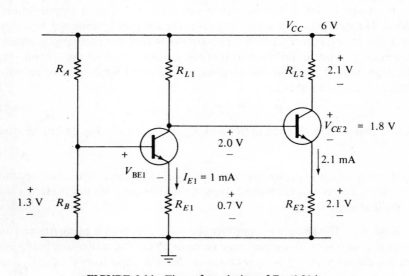

FIGURE 6-14 Figure for solution of Ex. 6-2(a).

across R_B is

$$V_{RB} = \frac{R_B}{R_A + R_B} V_{CC} = \frac{3.3 \text{ k}\Omega}{3.3 \text{ k}\Omega + 12 \text{ k}\Omega} \times 6 \text{ V} = 1.3 \text{ V}$$

The voltage across R_{E1} is

$$V_{RE1} = V_{RB} - V_{BE1} = 1.3 - 0.6 = 0.7 \text{ V}$$

The current I_{E1} becomes

$$I_{E1} = \frac{V_{RE1}}{R_{E1}} = \frac{0.7 \text{ V}}{0.7 \text{ k}\Omega} = 1 \text{mA}$$

and
$$V_{CE1} = V_{CC} - I_{E1}(R_{L1} + R_{E1})$$
$$= 6.0 - 1 \times 10^{-3}(4 \times 10^3) = 2 \text{ V}$$

Similarly, I_{E2} becomes

$$I_{E2} = \frac{V_{CE1} + V_{RE1} - V_{BE2}}{R_{E2}} = \frac{2.1 \text{ V}}{1 \text{ k}\Omega} = 2.1 \text{ mA}$$

(b) The final step is to divide the problem into a series of smaller problems that contain circuits whose properties are well known. This procedure is illustrated in Fig. 6-15. The voltage gain of the first stage can now be found,

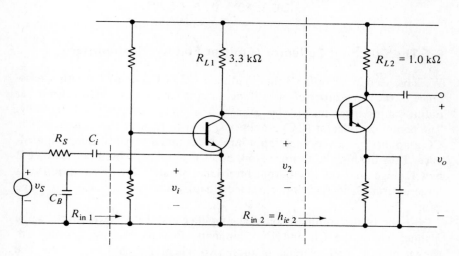

FIGURE 6-15 Figure for solution of Ex. 6-2(b).

and is given by v_2/v_i. We must remember to include the loading effect of the second stage:

$$R_L = R_{L1} \| h_{ie2}$$

$$h_{ie2} = \beta_2 r_{e2} + r_{b2} = \left(50 \times \frac{26}{2.1}\right) + 20 \simeq 630 \text{ } \Omega$$

$$\frac{v_2}{v_i} = +\frac{R_L}{r_{e1}} = \frac{3.3 \text{ k}\Omega \parallel 630 \text{ }\Omega}{26 \text{ }\Omega} = \frac{528}{26} = 203$$

where
$$r_{e1} = \frac{mv_T}{I_{E1}} = \frac{0.026 \text{ V}}{1 \text{ mA}} = 26 \text{ }\Omega$$

The voltage gain of the second stage is given by v_o/v_2 and is found from

$$\frac{v_o}{v_2} = -\frac{R_{L2}}{r_{e2}} = \frac{1 \text{ k}\Omega}{12.4 \text{ }\Omega} \simeq 80$$

where
$$r_{e2} = \frac{mv_T}{I_{E2}} = \frac{0.026 \text{ V}}{2.1 \text{ mA}} = 12.4 \text{ }\Omega$$

The voltage gain (a gain of less than unity) due to the source resistance is given by v_i/v_s and solved to be

$$\frac{v_i}{v_s} = \frac{R_{in1}}{R_s + R_{in1}} = \frac{26}{26 + 25} = 0.51$$

where R_{in1} is the input impedance of the first stage—a CB amplifier. Hence the overall voltage gain becomes the product of these three gains:

$$\frac{v_o}{v_s} = \frac{v_o}{v_2} \frac{v_2}{v_i} \frac{v_i}{v_s}$$
$$= 203 \times 80 \times 0.51 = 8{,}300$$

6.4 The Common Collector (Emitter Follower) Amplifier

We have seen in Section 6.3 that loading a CE or CB amplifier with a resistance of the same order of magnitude, or smaller than, the load resistor (or output impedance) makes the gain of that stage decrease substantially. The common collector (CC) amplifier is used to remedy this situation. The CC amplifier has a very high input impedance and a very low output impedance. Thus the high input impedance of a CC amplifier would not load a CB or CE stage, and the low output impedance would allow the addition of a large external load (i.e., low resistance load) without a decrease in voltage gain.

Biasing of the common collector amplifier may be achieved in the same manner as for the CE and CB amplifiers. Possible biasing schemes are shown in Fig. 6-16. Neither of these two schemes is used very often. The circuit in Fig. 6-16a is not used because the Q-point is not stable with temperature; and the circuit in Fig. 6-16b does not have a high enough input impedance to be practical. Even if R_{in} in Fig. 6-16b were high enough, the input impedance would be limited by the biasing resistors R_A and R_B, which are in parallel at the input. Hence biasing is usually achieved by direct coupling to the previous stage, as shown in Fig. 6-17.

If we consider the CC amplifier alone, as shown in Fig. 6-18, we can ex-

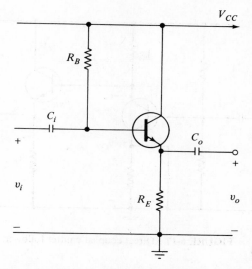

(a) Constant-current biased common collector amplifier

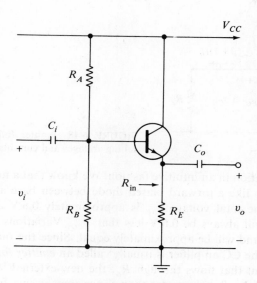

(b) H-type common collector amplifier

FIGURE 6-16 Biasing schemes for CC amplifier.

130   Chap. 6 / Common Base and Common Collector Amplifiers

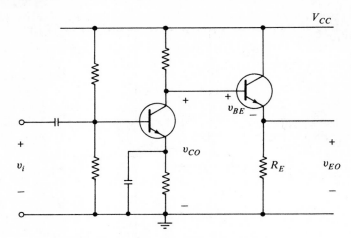

FIGURE 6-17   Direct coupled emitter follower.

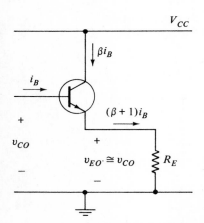

FIGURE 6-18   Emitter follower with the resulting voltages and currents.

plain its operation in an intuitive fashion. We know that a normally operated transistor acts like a forward biased diode between base and emitter. This means that the total voltage v_{BE} is approximately 0.6 V at all times; and voltage v_{EO} will always be 0.6 V less than v_{CO}. Variations at v_{CO} will also occur at v_{EO} and will be approximately equal. Since the output voltage v_{EO} follows v_{CO}, the CC amplifier is usually called an *emitter follower*.

The current that flows through R_E, the new external load, as shown in Fig. 6-18, will be $\beta + 1$ greater than the current drawn from the previous stage. Hence we have only current amplification with approximately unity voltage gain.

To add to our knowledge of the emitter follower, the input impedance, output impedance, and voltage gain should be calculated, using small-signal equivalent circuits. We shall do this using the *h*-parameter equivalent circuit for the transistor. Figure 6-19 shows the emitter follower and its complete equivalent circuit. In Fig. 6-19b R_G and v_s represent the Thévenin equivalent of the previous stage.

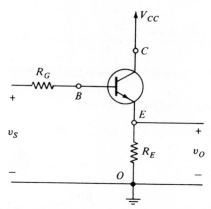

(a) The emitter follower with source resistance.

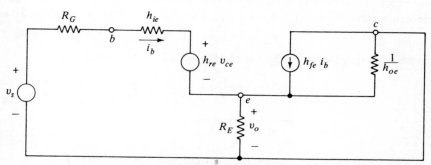

(b) The ac equivalent circuit of a common collector emitter follower with source resistance.

FIGURE 6-19 Emitter follower.

The same type of approximations used previously are now used to simplify the equivalent circuit in Fig. 6-19b. The voltage $h_{re}v_{ce}$ is neglected, and $1/h_{oe}$ is either included with R_E, or neglected if $1/h_{oe}$ is much greater than R_E. These approximations lead to the simplified CC equivalent circuit shown in Fig. 6-20.

To calculate the input impedance into the emitter follower, we obtain the ratio v_{bo}/i_b. From a statement of KVL, and Fig. 6-20, we can write

$$v_{bo} = h_{ie}i_b + (h_{fe} + 1)i_b R_E \qquad (6\text{-}18)$$

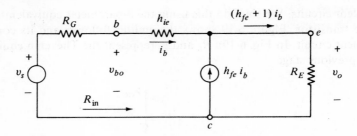

FIGURE 6-20 The simplified equivalent circuit for the emitter follower.

so that R_{in}, the input impedance, becomes

$$R_{in} = \frac{v_{bo}}{i_b} = h_{ie} + (h_{fe} + 1)R_E$$
$$\simeq (h_{fe} + 1)R_E \qquad \text{if } (h_{fe} + 1)R_E > h_{ie} \tag{6-19}$$

R_{in} can also be written as

$$R_{in} \simeq \beta R_E \tag{6-20}$$

Next we calculate the voltage gain. Since we are likely to include the loading effect of the emitter follower (usually negligible) on the previous stage, we can define the voltage A_v as v_o/v_{bo}. The output voltage is calculated as

$$v_o = +(h_{fe} + 1)i_b R_E \tag{6-21}$$

and by dividing Eq. (6-21) by Eq. (6-18), we obtain the voltage gain

$$A_v = \frac{v_o}{v_{bo}} = \frac{(h_{fe} + 1)i_b R_E}{h_{ie}i_b + (h_{fe} + 1)i_b R_E}$$
$$= \frac{R_E}{R_E + \dfrac{h_{ie}}{h_{fe} + 1}} \tag{6-22}$$

This expression can be rewritten in terms of r-parameters as

$$A_v = \frac{R_E}{R_E + r_e} \simeq 1 \tag{6-23}$$

since $r_e \ll R_E$ for most cases. Notice that the voltage gain is always less than unity but, for most cases, can be approximated at unity.

To obtain the output impedance of the emitter follower, the Thévenin equivalent of Fig. 6-20 must be found. This procedure is not quite as simple as for the previous amplifier configurations that we have studied.

First, we shall need the open-circuit voltage v_o:

$$v_o = (h_{fe} + 1)i_b R_E \tag{6-24}$$

The input voltage is given as

$$v_s = i_b(R_G + h_{ie}) + (h_{fe} + 1)i_b R_E \tag{6-25}$$

Substituting i_b, obtained from Eq. (6-25), into Eq. (6-24), v_o ($= v_{oc} =$ open-circuit voltage) becomes

$$v_{oc} = v_o = \frac{(h_{fe} + 1)R_E}{(R_G + h_{ie}) + (h_{fe} + 1)R_E} v_s \qquad (6\text{-}26)$$

We must be careful when obtaining the short-circuit current, because the short-circuit changes the value of the base current. We shall call the new base current i_b'. See Fig. 6-21. Now,

$$i_b' = \frac{v_s}{R_G + h_{ie}} \qquad (6\text{-}27)$$

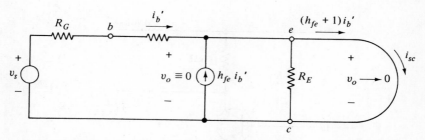

FIGURE 6-21 The emitter follower equivalent circuit under a short circuit condition.

and the output short-circuit current i_{sc} is

$$i_{sc} = (h_{fe} + 1)i_b' \qquad (6\text{-}28)$$

Substituting the value of i_b' into Eq. (6-28), we obtain

$$i_{sc} = (h_{fe} + 1)\frac{v_s}{R_G + h_{ie}} \qquad (6\text{-}29)$$

To obtain the output impedance R_{out}, we divide Eq. (6-26) by Eq. (6-29):

$$\begin{aligned} R_{out} &= \frac{v_{oc}}{i_{sc}} \\ &= \frac{(h_{fe} + 1)R_E}{[(R_G + h_{ie}) + (h_{fe} + 1)R_E]} \frac{R_G + h_{ie}}{h_{fe} + 1} \\ &\simeq \frac{\frac{R_G + h_{ie}}{h_{fe}} R_E}{\frac{R_G + h_{ie}}{h_{fe}} + R_E} \\ &= R_E \parallel \left(\frac{R_G + h_{ie}}{h_{fe}}\right) \end{aligned} \qquad (6\text{-}30)$$

If we substitute, in terms of the equivalent r-parameters, we obtain

$$R_{out} = R_E \parallel \left(\frac{R_G}{\beta} + r_e\right) \qquad (6\text{-}31)$$

Equation (6-31) can be interpreted in the following fashion. The source resistance contribution is decreased by the factor β, to which a small contribution from r_e is added. To this result, the resistor R_E is added in parallel. In many cases, R_E is much larger than $(R_G/\beta) + r_e$, so Eq. (6-31) becomes

$$R_{out} \simeq \left(\frac{R_G}{\beta} + r_e\right) \tag{6-32}$$

Example 6-3:
Given:
(a) The amplifier circuit shown in Fig. 6-22.

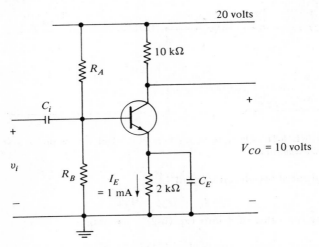

FIGURE 6-22 Problem statement for Ex. 6-3.

(b) A supply of silicon *npn* transistors whose β's are approximately 120.
Required:
(a) To add an emitter follower to decrease the output impedance and maintain the voltage gain of the given amplifier.
(b) To calculate the resulting output impedance.

Solution:

(a) The addition of an emitter follower is shown in Fig. 6-23. The choice of R_{E2} depends only on the amount of emitter current that is desired by the designer. If 2 mA is chosen, R_{E2} becomes

$$R_{E2} = \frac{9.4 \text{ V}}{2 \text{mA}} = 4.7 \text{ k}\Omega$$

The voltage gain is maintained, since the input impedance becomes

$$R_{in} = h_{ie} + (h_{fe} + 1)R_{E2}$$
$$\simeq (120)(4.7 \text{ k}\Omega) = 560 \text{ k}\Omega$$

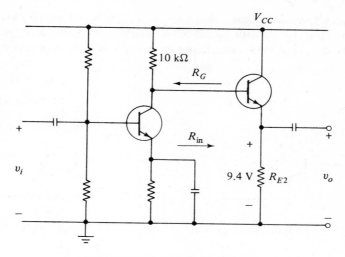

FIGURE 6-23 Solution for Ex. 6-3.

This input impedance is certainly much larger than the output impedance of the first stage; hence the first stage will not be loaded.

(b) The output impedance of the emitter follower is

$$R_{out} = R_{E2} \parallel \left(\frac{R_G}{\beta} + r_e\right)$$
$$= 4.7 \text{ k}\Omega \parallel \left(\frac{10 \text{ k}\Omega}{120} + 13 \, \Omega\right) = 47 \text{ k}\Omega \parallel (83 + 13) \, \Omega$$
$$\simeq 83 + 13 = 96 \, \Omega$$

Notice that the effect of R_E is negligible for the calculation of the output impedance in Example 6-3, as implied in Eq. (6-32).

6.5 Load Line Analysis for the Emitter Follower

We previously used, as biasing criterion for the CE amplifier, the following rule: The voltage v_{CE} should be slightly under one-half V_{CC} to obtain the maximum output voltage swing. This is not always true, particularly for the emitter follower. The ac and the dc loads are so different in the CC case, especially when an external load is applied. In the case of an emitter follower, the designer may wish to use a very large external load (i.e., low resistance), because the output impedance is so much lower than for a CB or CE amplifier.

Load line analysis for the CC is identical to that for the CE amplifier, except that R_L (resistor between the collector and the supply) is equal to zero. The method of constructing load lines was previously discussed in Section 4.6.

Figures 6-24 and 6-25 illustrate how a choice of Q-point can affect the maximum allowable output voltage without clipping. In Fig. 6-24a, the Q-point has been chosen with V_{CE} at 5.0 V for the given value of V_{CC} equal to 10 V. The dc and ac load lines are shown in Fig. 6-25. The amplifier has been

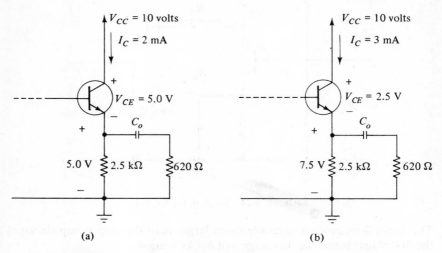

FIGURE 6-24 An emitter follower biased at two Q points. (a) $V_{CE} = 5.0$ volts, and $I_C = 2$ ma. (b) $V_{CE} = 2.5$ volts, and $I_C = 3$ ma.

loaded down with 620 Ω, so that the effective load resistance is 500 Ω. The resulting output voltage swing v_{ce1} can vary 2 V peak to peak (p-p) under this condition. In Fig. 6-24b, the same amplifier is used again, but the biasing voltage V_{CE} has been decreased to 2.5 V. This establishes a new Q-point, Q_2, shown in Fig. 6-25. The ac load is now drawn (shown as the broken line) through Q_2 with the same slope as before, since the same load is being used. Notice that at this new Q-point the resulting output voltage v_{ce2} can now swing by 3 V, peak to peak.

It is apparent from this example that the designer should be very careful about the choice of biasing for the emitter follower, unless, of course, he is not concerned with large output voltages. Usually, a quick sketch of the resulting load lines during the design helps to provide a better insight into the problem, thus avoiding a change when the circuit has been built.

6.6 Bootstrapped Amplifiers

Both the unbypassed emitter resistor amplifier and the emitter follower offer high input impedance when they are biased by the previous stage. If either of these two amplifiers is used by itself, or as the first stage of a multistage amplifier, the input impedance is limited by the biasing resistors R_A

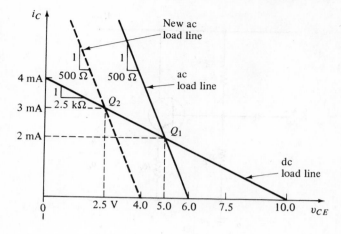

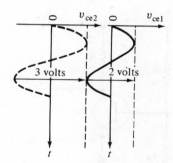

FIGURE 6-25 Load line analysis for a *CC* amplifier biased at two different *Q* points.

and R_B. The input impedance of the amplifiers, shown in Fig. 6-26 is given by

$$R_{in} = R_A \| R_B \| (\beta + 1)R_E$$

and even though $(\beta + 1)R_E$ is quite high, the parallel combination of R_A and R_B reduces R_{in}, so that it may be only a few times larger than for a CE amplifier. To achieve higher input impedance and to overcome this difficulty, the bootstrap circuit has been developed. With this circuit, it is possible to obtain an input impedance in excess of 250 kΩ. This becomes a very useful circuit when an amplifier is to be driven from a high impedance source. Figure 6-27 illustrates two bootstrapped circuits.

Figure 6-27a is a low-gain amplifier with high input impedance, using an unbypassed emitter resistor, and Fig. 6-27b is a bootstrapped emitter follower. We shall analyze the first circuit in Fig. 6-27a using *h*-parameters and leave the second as an exercise for the reader. Analysis is practically the same. The

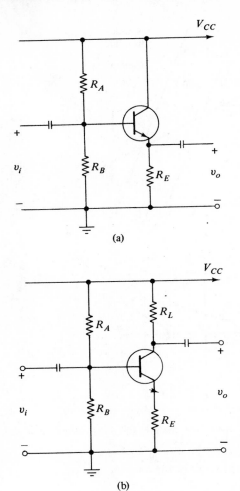

(a) The H-type biased emitter follower, and
(b) The H-type biased amplifier with an unbypassed emitter resistor

FIGURE 6-26 Two amplifier circuits with unbypassed emitter resistors.

equivalent circuit for this amplifier is shown in Fig. 6-28. All the capacitors in the circuit are assumed to be short circuits to ac, and the effect of h_{re} and h_{oe} is neglected. Notice that the purpose of C_B is to place R_E, R_A, and R_B between the emitter and ground (in a small-signal sense). R_c is effectively situated between base and emitter, or in parallel with h_{ie}. Let the current through R_c be designated i_x and the parallel combination of R_A, R_B, and R_E be R_F, that is, $R_F = R_A \| R_B \| R_E$.

Sec. 6.6 / Bootstrapped Amplifiers

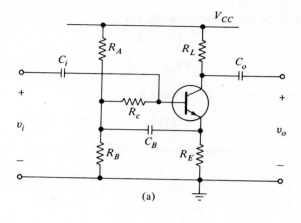

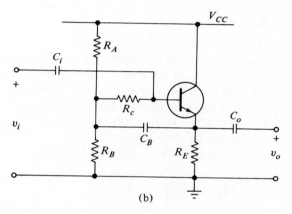

FIGURE 6-27 Bootstrapped amplifiers: (a) the unbypassed emitter resistor amplifier, (b) the emitter follower.

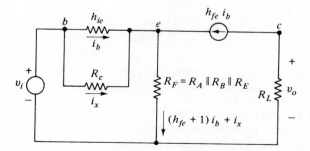

FIGURE 6-28 The equivalent circuit for a bootstrapped amplifier with an unbypassed emitter resistor.

From Fig. 6-28, we can write, using KVL,
$$h_{ie}i_b = R_c i_x \tag{6-33}$$
so that
$$i_x = \frac{h_{ie}}{R_c} i_b \tag{6-34}$$

Using KVL at the input, we express v_i as
$$v_i = h_{ie}i_b + R_F\left(h_{fe} + 1 + \frac{h_{ie}}{R_c}\right) i_b \tag{6-35}$$

To find the input impedance, we divide Eq. (6-35) by the input current $i_b + i_x$:

$$R_{in} = \frac{v_i}{i_b + i_x} = \frac{v_i}{i_b\left(1 + \frac{h_{ie}}{R_c}\right)} = \frac{h_{ie} + \left(h_{fe} + \frac{h_{ie}}{R_c}\right)R_F}{1 + \frac{h_{ie}}{R_c}} \tag{6-36}$$

From Eq. (6-36) it is apparent that if h_{ie}/R_c is a small in comparison to unity, then R_{in} is reduced to
$$R_{in} \cong h_{ie} + (h_{fe} + 1)R_F \qquad \text{for } R_c > h_{ie} \tag{6-37}$$
Usually R_c can be made about 20 kΩ before the dc voltage drop across it ruins the temperature stability of the H-type biasing circuit. The dc voltage drop across R_c should be small in comparison to the voltage across R_E.

From Eq. (6-37), it is apparent that input impedance can be made quite high. It is interesting to note the similarity between Eq. (6-37) and Eq. (6-19) for the input impedance for an emitter follower.

To obtain the voltage gain, we find the output voltage v_o
$$v_o = -h_{fe}i_b R_L \tag{6-38}$$
and divide the Eq. (6-38) by Eq. (6-35):

$$A_v = \frac{v_o}{v_i} = -\frac{h_{fe}R_L}{h_{ie} + R_F\left(h_{fe} + 1 + \frac{h_{ie}}{R_c}\right)} \tag{6-39}$$

$$\simeq -\frac{R_L}{R_F} \quad \begin{cases} \text{if } \frac{h_{ie}}{R_c} < h_{fe} \\ \text{and } h_{ie} < R_F h_{fe} \end{cases}$$

The expression for voltage gain in Eq. (6-39) has the same form as the unbypassed emitter resistor amplifier. In general, R_F will be smaller than R_E, and so, a gain of 2.0 to about 10.0 is quite easy to obtain.

Example 6-4: Given a *pnp* silicon transistor with $h_{fe} \simeq 100$, and a 10-V dc supply, design a bootstrapped amplifier with an input impedance in excess of 75 kΩ and a voltage gain in excess of 5.0.

Solution:

(a) The first step is to choose approximate values for biasing, keeping Eqs. (6-37) and (6-39) in mind. This is done in Fig. 6-29. R_E is chosen as 2 kΩ

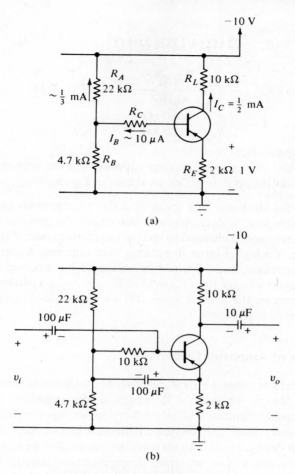

FIGURE 6-29 Circuits for Ex. 6-4.

to meet the input impedance condition, and R_L is chosen as 10 kΩ to meet the gain specification, remembering that R_F is less than R_E. If R_c is chosen as 10 kΩ, the voltage drop will be about 0.1 V, and negligible when compared with the 1.0 V across R_E. If h_{fe} should change under temperature, the voltage across R_c will not seriously affect the bias voltages. R_A and R_B are then calculated in the usual way.

(b) The next step is to calculate the voltage gain and input impedance and see if the estimated values will meet the specifications.

$$R_{in} = \frac{h_{ie} + \left(h_{fe} + \dfrac{h_{ie}}{R_c}\right) R_F}{1 + \dfrac{h_{ie}}{R_c}}$$

where

$$R_F = 22 \text{ k}\Omega \| 4.7 \text{ k}\Omega \| 2.0 \text{ k}\Omega$$
$$\simeq 1.3 \text{ k}\Omega$$
$$R_{in} = \frac{5.2 \text{ k}\Omega + (100 + 0.5)1.3 \text{ k}\Omega}{1.5} \simeq 90 \text{ k}\Omega$$
$$A_v = -\frac{R_L}{R_F} = -\frac{10 \text{ k}\Omega}{1.3 \text{ k}\Omega} = -7.7$$

Both requirements have been met.

(c) The final step is to place large capacitors in the appropriate places to obtain a bootstrapped amplifier, as shown in Fig. 6-29b.

Example 6-4 illustrates how specified design requirements can be met in an approximate way. Such calculations also enable the designer to see those parameters that may be changed to change the performance of the amplifier. For example, if a higher input impedance were required, R_E and R_B would have to be increased, and h_{ie} would be decreased by choosing a higher I_C. This may require a larger V_{CC} or a smaller R_L. Choosing a smaller R_L would, of course, decrease the gain. A series of such calculations could be used if specific criteria are to be met.

6.7 Design of Amplifier Circuits

In the previous sections we have covered the fundamental building blocks of amplifier design. To successfully design simple amplifiers, the designer must be thoroughly familiar with all the formulas for input impedance, output impedance, and voltage gain for these individual blocks. After this has been achieved, the designer is able to plan his circuits as soon as he is introduced to the specifications. For example, if specifications require an amplifier with a gain of several hundred, a low input impedance, and a low output impedance, the designer would immediately decide on a two-stage amplifier: a CB followed by a CC stage.

Table 6-1 contains a summary of all the pertinent formulas. It would be well for the student to memorize these and know how to use them for designing one-, two-, or three-stage amplifiers.

6.8 Some Examples of Unusual Cascading

The term *cascading* is used to describe the process of connecting two amplifier stages. We have already used two methods of cascading stages. The first was the use of a coupling capacitor C_c to connect the two stages. This method is called *RC coupling*, the name implying the elements used to cascade the

TABLE 6-1 SUMMARY OF AMPLIFIERS AND PERTINENT FORMULAS

Amplifier Stage	Input Impedance R_{in}		Voltage Gain A_v		Output Impedance R_{out}	
	h	r	h	r	h	r
Common emitter CE	$R_A \| R_B \| h_{ie}$	$R_A \| R_B \| (r_b + (\beta+1)r_e)$	$-\dfrac{h_{fe} R_L}{h_{ie}}$	$-\dfrac{\beta R_L}{r_b + (\beta+1)r_e}$ $\simeq -\dfrac{R_L}{r_e}$	R_L	R_L
Common collector CC	$h_{ie} + (h_{fe}+1)R_E$ $\simeq h_{fe} R_E$	$r_b + (\beta+1)(r_e + R_E)$ $\simeq \beta R_E$	$\simeq 1$	$\simeq 1$	$R_E \| \dfrac{(h_{ie}+R_G)}{h_{fe}}$ $\simeq \dfrac{h_{ie}+R_G}{h_{fe}}$	$R_E \| \left(r_e + \dfrac{R_G}{\beta}\right)$ $\simeq r_e + \dfrac{R_G}{\beta}$
Common base CB	$R_E \| \dfrac{h_{ie}}{h_{fe}+1}$ $\simeq \dfrac{h_{ie}}{h_{fe}}$	$R_E \| r_e$ $\simeq r_e$	$+\dfrac{h_{fe} R_L}{h_{ie}}$	$+\dfrac{\beta R_L}{r_b + (\beta+1)r_e}$ $= +\dfrac{R_L}{r_e}$	R_L	R_L
Unbypassed emitter resistor amplifier	$R_A \| R_B \| (h_{fe}+1) R_E$ $= (h_{fe}+1) R_E$ *	$R_A \| R_B \| (\beta+1) R_E$ $= (\beta+1) R_E$ *	$\dfrac{-h_{fe} R_L}{h_{ie} + (h_{fe}+1) R_E}$ $\simeq -\dfrac{R_L}{R_E}$	$\dfrac{-\beta R_E}{r_b + (\beta+1)(r_e + R_E)}$ $\simeq -\dfrac{R_L}{R_E}$	R_L	R_L

*This is the formula when R_A and R_B are not used.

stages. The second method was direct coupling. Figure 6-30 illustrates two CE amplifiers cascaded by both methods.

The individual stages in a circuit such as they are shown in Fig. 6-30 are easy to recognize and can be easily analyzed by the use of appropriate formulas from Table 6-1. However, sometimes the individual stages in a cascaded amplifier are not so easily recognized. Such is the case with the cascode amplifier shown in Fig. 6-31.

After careful observations, it is easy to see that the cascode amplifier is made up of two direct coupled amplifier stages. The dashed line in Fig. 6-31

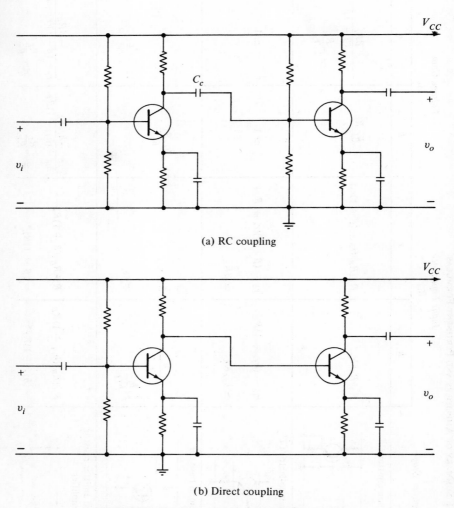

FIGURE 6-30 Two examples of amplifier coupling.

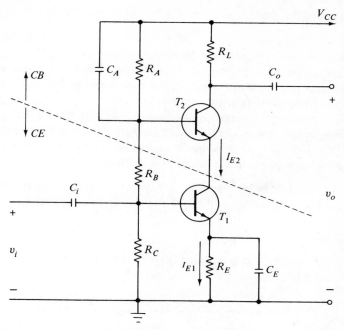

FIGURE 6-31 The cascode amplifier as an example of unusual coupling.

indicates the individual stages, with T_1 acting as a CE amplifier directly coupled to T_2, the CB amplifier. C_A bypasses R_A to make T_2 a CB stage. Notice that the input impedance of T_2, r_{e2}, is the load for T_1. Hence the gain of the first stage v_{ce1}/v_i can be calculated:

$$\frac{v_{ce1}}{v_i} = -\frac{r_{e2}}{r_{e1}} \tag{6-40}$$

Since $I_{E1} \simeq I_{E2}$ Eq. (6-40) reduces to

$$\frac{v_{ce1}}{v_i} = -\frac{\dfrac{v_T}{I_{E2}}}{\dfrac{v_T}{I_{E1}}} \simeq -1 \tag{6-41}$$

The gain of the second stage can be calculated as

$$\frac{v_o}{v_{ce1}} = +\frac{R_L}{r_{e2}} \tag{6-42}$$

The overall gain is the product of Eqs. (6-41) and (6-42):

$$\frac{v_o}{v_i} = A_v = -\frac{R_L}{r_{e2}} \tag{6-43}$$

The input impedance can be calculated as

$$R_{in} = R_B \parallel R_C \parallel h_{ie1} \tag{6-44}$$

where h_{ie1} is the parameter h_{ie} of T_1. The output impedance is equal to R_L,

$$R_{out} = R_L \qquad (6\text{-}45)$$

The electronic properties that we have observed (R_{in}, R_{out}, and A_v) are the same as for the CE amplifier. The cascode amplifier, therefore, has the additional property of operation over a wide range of frequencies. We shall examine this property in a later chapter. It is interesting to observe the method used to determine the properties of the cascode amplifier. The amplifier was split into simple circuits and the equations from Table 6-1 were employed. This is much easier than drawing the equivalent circuit for the whole amplifier and trying to analyze it according to Kirchhoff's voltage and current laws.

Example 6-5: Given two *npn* silicon transistors with $h_{fe} \simeq 100$ and a 12-V supply, bias the cascode amplifier shown in Fig. 6-31 and calculate the resulting gain.

Solution:

(a) The circuit is redrawn in Fig. 6-32, and appropriate voltages are assigned across R_L, R_E, and across the transistors.

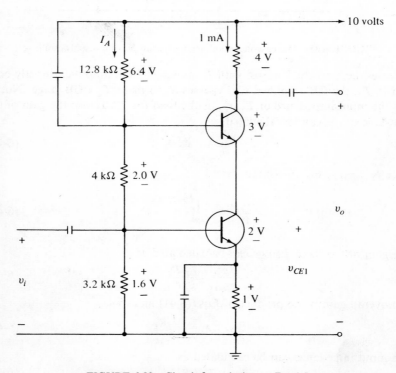

FIGURE 6-32 Circuit for solution to Ex. 6-5.

(b) A collector current of 1 mA is chosen to flow through both transistors and the values of R_L and R_E are calculated:

$$R_L = \frac{4\text{ V}}{1\text{ mA}} = 4\text{ k}\Omega$$

$$R_E = \frac{1\text{ V}}{1\text{ mA}} = 1\text{ k}\Omega$$

(c) The voltages across R_A, R_B, and R_C are found by remembering that the base emitter voltage should be 0.6 V for both silicon transistors.

(d) A current I_A, which is much larger than the base currents in the transistors, is chosen to be 0.5 mA. The resistors R_A, R_B, and R_C are now calculated to be 12.8, 4, and 3.2 kΩ, respectively.

(e) The voltage gain for this amplifier is

$$A_v \simeq -\frac{R_L}{r_{e2}} \simeq -\frac{4\text{ k}\Omega}{26\Omega} = -154$$

Another example of unusual cascading of amplifier stages is the difference amplifier, shown in Fig. 6-33. The biasing in this circuit is supplied by two batteries rather than a voltage divider, as in the H-type biasing.

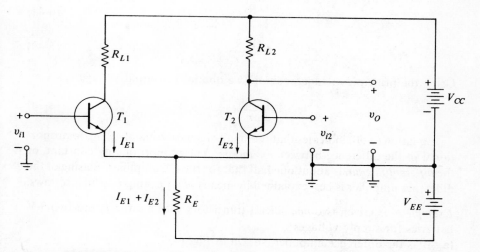

FIGURE 6-33 The difference amplifier as an example of unusual cascading of amplifiers.

The operation of the difference amplifier can be explained in the following way, using the principle of superposition. If v_{i2} is set to zero, then T_1 acts as a common collector amplifier and T_2 acts as a CB amplifier. The output impedance of T_1 is r_{e1}, and the input impedance of T_2 is r_{e2}, since it is a CB amplifier. Hence the CB amplifier is driven by a source whose output impedance is r_{e1}. If the same emitter current is used in both stages, then the gain v_o/v_{i1} is given by

$$A_{v1} = \left.\frac{v_o}{v_{i1}}\right|_{v_{i2}=0} = \left(\frac{\beta R_{L2}}{r_{b2} + (\beta+1)r_{e2}}\right)\left(\frac{r_{e1}}{r_{e1}+r_{e2}}\right) \quad (6\text{-}46)$$

$$\simeq \frac{1}{2}\frac{R_{L2}}{r_e}$$

where $\qquad r_e = r_{e1} = r_{e2} \quad$ since $I_{E1} = I_{E2}$

If v_{i1} is now set equal to zero, then T_2 acts like a common emitter amplifier, with an unbypassed emitter resistor, with T_1 as a common base amplifier. The input impedance into T_1 is r_e, and this is essentially the ac resistance between the emitter of T_1 and the ground (unbypassed emitter resistor). Hence the gain of T_2 due to v_{i2} can be calculated as

$$A_{v2} = \left.\frac{v_o}{v_{i2}}\right|_{v_{i1}=0} = -\frac{\beta R_L}{r_b + (\beta+1)(r_{e2}+r_{e1})} \quad (6\text{-}47)$$

$$\simeq -\frac{1}{2}\frac{R_{L2}}{r_e}$$

Equations (6-46) and (6-47) can be rewritten as

$$v_o = \frac{1}{2}\frac{R_{L2}}{r_e}v_{i1}\bigg|_{v_{i2}=0} \quad (6\text{-}48)$$

$$v_o = -\frac{1}{2}\frac{R_{L2}}{r_e}v_{i2}\bigg|_{v_{i1}=0} \quad (6\text{-}49)$$

Using the principle of superposition, we obtain the output voltage v_o:

$$v_o = \frac{1}{2}\frac{R_{L2}}{r_e}(v_{i1} - v_{i2}) \quad (6\text{-}50)$$

Equation (6-50) indicates that the small-signal portion of v_o, v_o, is proportional to the difference between v_{i1} and v_{i2}. The proportionality constant, or the *differential gain*, is about one half that for the CE amplifier. Biasing of the difference amplifier is best explained by means of an example, which follows.

Example 6-6: Given two *npn* silicon transistors with a large β and two 9-V batteries for supply voltages:
(a) Bias the difference amplifier, shown in Fig. 6-33.
(b) Calculate the resulting differential gain.

Solution:

(a) The circuit is redrawn in Fig. 6-34 for biasing purpose. Both inputs are set to zero by grounding them.
(b) Identical load resistors and collector currents are chosen for both transistors of 2 kΩ and 2 mA, respectively. Since the emitter voltage is 0.6 V below the base potential, V_{CE} is equal at 5.6 V for both transistors.

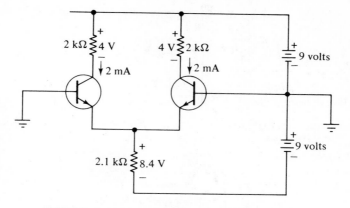

FIGURE 6-34 The circuit for the solution of Ex. 6-6.

(c) The voltage across R_E is now

$$V_{RE} = V_{EE} - V_{BE} = -8.4 \text{ V}$$

R_E can now be calculated, since the current through R_E is 4.0 mA, the sum of the two emitter currents:

$$R_E = \frac{8.4 \text{ V}}{4\text{mA}} = 2.1 \text{ k}\Omega$$

(d) No coupling capacitors are necessary at the inputs, since the bases are already at zero potential. The input voltages are applied directly between the bases and the ground. An output coupling capacitor may be used if the dc level must be removed from the output; otherwise it is not required (see Fig. 6-33).

(e) The differential voltage gain is calculated as

$$A_{vd} = \frac{1}{2} \frac{R_{L2}}{r_e} \simeq \frac{1}{2} \frac{2 \text{ k}\Omega}{13 \text{ }\Omega} \simeq 77$$

Another final example of unusual coupling is the *Darlington connection,* shown in Fig. 6-35. This circuit is not an amplifier circuit, like the circuits discussed previously but can be used to make up an amplifier circuit such as the CB, CE, and CC stages. In brief, the Darlington connection is the manner of connecting two transistors, as shown in Fig. 6-35a. which may be considered to act as a single transistor with an overall current gain approximately the product of the individual current gains. To illustrate this, we note that the overall collector current i_C of the transistors in Fig. 6-35a, acting as a single transistor, is the sum of the two individual collector currents i_{C1} and i_{C2}:

$$i_C = i_{C1} + i_{C2}$$

If the current gains of T_1 and T_2 are h_{FE1} and h_{FE2}, respectively, Eq. (6-51)

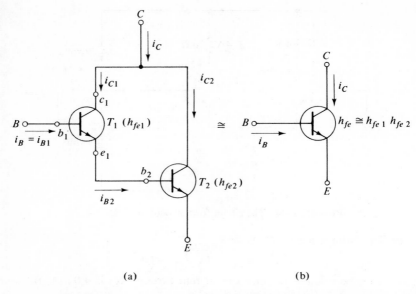

FIGURE 6-35 The Darlington connection: (a) Two transistors with gains h_{fe}, and h_{fe} connected in the Darlington configuration. (b) The single transistor equivalent of the Darlington connection.

can be written as

$$i_C = h_{FE1}i_{B1} + h_{FE2}i_{B2} \tag{6-52}$$

but since
$$i_{B2} = (h_{FE1} + 1)i_{B1} \tag{6-53}$$

and
$$i_B = i_{B1}$$

then
$$i_C = [h_{FE1} + (h_{FE1} + 1)h_{FE2}]i_B \tag{6-54}$$

Equation (6-54) can now be approximated as follows:

$$i_C \simeq h_{FE1}h_{FE2}i_B \tag{6-55}$$

Thus the overall current gain of the Darlington connection h_{FE} is given by

$$h_{FE} = \frac{i_C}{i_B} \simeq h_{FE1}h_{FE2} \tag{6-56}$$

It is easy to see that the small-signal overall current gain h_{fe} will be

$$h_{fe} \simeq h_{fe1}h_{fe2} \tag{6-57}$$

where h_{fe1} and h_{fe2} are small-signal current gains of T_1 and T_2, respectively.
It can also be shown that the overall short circuit input impedance h_{ie} is

$$h_{ie} = h_{ie1} + (h_{fe1} + 1)h_{ie2} \tag{6-58}$$

Equation (6-58) can be obtained by drawing the small-signal equivalent circuit for Fig. 6-35 and obtaining the short-circuit input impedance of the Darlington connected circuit. Equation (6-58) can also be verified by noticing that h_{ie2} acts like the emitter resistor to the first transistor in an emitter follower, in Fig. 6-35a. Thus, if we calculate the input impedance using the formula for an emitter follower consisting of T_1 and h_{ie2}, we obtain Eq. (6-58).

The Darlington connection finds use in circuits where either a large h_{ie} or h_{fe}, or possibly both, is needed to produce a desired result. Darlington-connected transistors find use in (1) difference amplifiers, where a large h_{ie} is needed to produce a large input impedance, and (2) emitter followers, where a large h_{fe} gives rise to a larger input impedance, for example.

Figure 6-36 illustrates the use of a Darlington connection in an amplifier

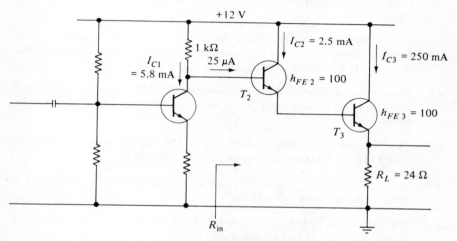

FIGURE 6-36 A partly biased amplifier illustrating the use of the Darlington connection in an emitter follower circuit.

employing a CE stage and in an emitter follower with a very small load resistance. Notice that R_{in}, the input impedance of the emitter follower, becomes

$$R_{in} = h_{fe}R_L + h_{ie}$$
$$\simeq h_{fe2}h_{fe3}R_L + h_{fe2}h_{ie3} + h_{ie2} \qquad (6\text{-}59)$$
$$\simeq h_{fe2}h_{fe3}R_L$$

PROBLEMS

6.1 Bias a common base amplifier at $I_E = 2$ mA and $V_{CE} = 5$ V. Use an *npn* germanium transistor.

6.2 A germanium transistor is biased in the common base configuration at $I_E = 3$ mA and $V_{CE} = 10$ V. The following parameters were measured for the transistor at this Q-point: $r_b = 200\,\Omega$, $\alpha = 0.98$, and $r_d = 1.0$ MΩ. Calculate the input impedance of the amplifier.

6.3 Design an amplifier with an input impedance of 52 Ω and a gain of 100. Use an *npn* silicon transistor.

6.4 If the amplifier of Problem 6.3 is connected to a signal generator with an open-circuit voltage of v_s and an output impedance of 52 Ω, calculate the overall gain v_o/v_s, where v_o is the output voltage.

6.5 Obtain the input impedance, output impedance, and voltage gain of a common base amplifier, using the CB equivalent tee model. Neglect the effect of r_c in your analysis.

6.6 Obtain the expressions for input impedance and output impedance of an emitter follower, using the CE equivalent tee model for a transistor. Make the usual type of approximations in your analysis.

6.7 Design an amplifier with input impedance of 52 Ω, an output impedance of 52 Ω, and a gain of 100. (*Hint:* Use a CB amplifier coupled to an emitter follower.)

6.8 Design a two-stage amplifier with $|A_v| = 150$, having an input impedance of 1.5 kΩ and an output impedance of less than 100 Ω.

6.9 An external load of 200 Ω is connected to the amplifier in Problem 6.8. What is the resulting overall gain?

6.10 Use a bootstrapped stage to design an amplifier with an input impedance in excess of 250 kΩ and a voltage gain of -2.

6.11 Use the simplified h-parameter equivalent circuit (using h_{ie} and h_{fe} only) to obtain the input impedance, voltage gain, and output impedance for the bootstrapped emitter follower shown in Fig. 6-27b. How do the results compare with the circuit in Fig. 6-27a?

6.12 (a) Analyze the circuit shown in Fig. P6-12 by explaining the purpose of each amplifier stage and then calculating the overall voltage gain, input impedance, and output impedance.
    (b) Bias the amplifier of Fig. P6-12.

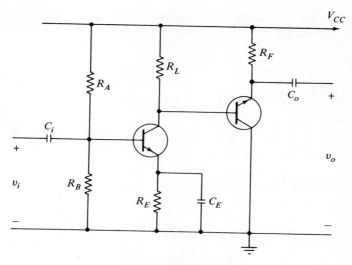

FIGURE P6-12

6.13 Bias and analyze the circuit shown in Fig. P6-13. All capacitors are large.

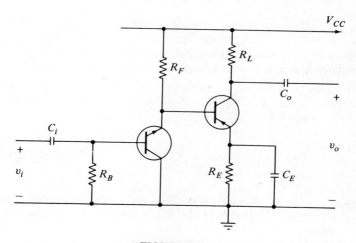

FIGURE P6-13

6-14. (a) Bias the amplifier shown in Fig. P6-14.
(b) Find R_{in}, R_{out}, and A_v for the circuit.
(c) What amplifier has similar properties to this circuit?

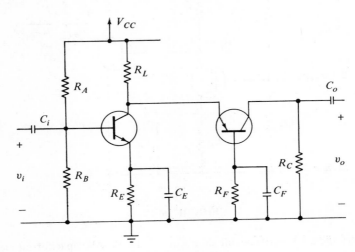

FIGURE P6-14

6.15 Design an amplifier to obtain an output voltage of 1 V peak to peak, when used with a crystal microphone. The crystal microphone has an open-circuit voltage of 0.05 V peak to peak and an output impedance of 0.5 MΩ. The amplifier is to have an output impedance of no more than 10 kΩ.

6.16 Design an amplifier to have a voltage gain of 100 when operated into a 100-Ω load. The amplifier is to be driven with a signal generator whose output impedance is 52 Ω.

REFERENCES

[1] SEARLE, C. L., BOOTHROYD, A. R., ANGELO, E. J., JR., GRAY, P. E., and PEDERSON, D. O., *Elementary Circuit Properties of Transistors*, Semiconductor Electronics Educational Committee, Vol. 3 (John Wiley & Sons, Inc., New York, 1964).

[2] FITCHEN, F. C., *Transistor Circuit Analysis and Design*, 2nd ed. (Van Nostrand Reinhold Co., Princeton, N.J., 1966).

[3] CHIRLIAN, P. M., *Analysis and Design of Electronic Circuits* (McGraw-Hill Book Co., New York, 1965).

[4] SCHILLING, D. L., and BELOVE, C., *Electronic Circuits: Discrete and Integrated* (McGraw-Hill Book Co., New York, 1968).

[5] SEELY, S., *Electronics Circuits* (Holt, Rinehart and Winston, Inc., New York, 1968).

7

THE FIELD EFFECT TRANSISTOR

The advent of solid-state technology has brought forth a large variety of solid-state devices. One of these devices is the *field effect transistor (FET)*. Although the impact of the FET was not felt on the electronics industry until about a decade after the development of the transistor, the FET is actually a very "old" device. The first patent granted for the FET dates back to 1928.*

The major difference between the ordinary transistor and FET is the very high input impedance of the FET. The ordinary transistor is usually called the *bipolar junction transistor (BJT)* to distingush it from the FET. The FET is manufactured in two different ways: the junction field effect (JFET) and the insulated gate effect (IGFET). The IGFET is sometimes also referred to as the metal oxide semiconductor field effect (MOSFET). The difference between the JFET and the IGFET is that the IGFET has a higher input impedance than the JFET.

7.1 Operation of the Junction FET (JFET)

A pictorial representation of the JFET is shown in Fig. 7-1a and the symbol in Fig. 7-1b. The FET is a three-terminal device made of a bar of *n*-type silicon material with a "ring" of *p*-type material at the center. The source (S) and drain (D) terminals are situated at each end of the *n*-type bar and the

---

*U.S. Patent 1900018 was granted to Lilienfeld for "a device controlling electric current."

156   Chap. 7 / The Field Effect Transistor

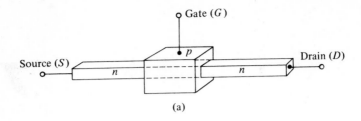

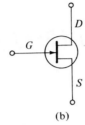

(a) A pictorial representation of the structure of an *n*-channel FET
(b) The corresponding symbol for an *n*-channel JFET.

FIGURE 7-1   The JFET.

gate (G) terminal is connected to the *p*-type material. This type of FET is referred to as the *n*-channel FET. The channel is the bar of *n*-type material to which the source and drain terminals are connected. *p*-channel FETs are made as well, with the gate of *n*-type material. The symbol for a *p*-channel FET is shown in Fig. 7-2.

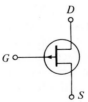

FIGURE 7-2   The symbol for a *p*-channel JFET.

Under normal operating conditions the FET is biased so that the gate–channel junction is reverse biased. This gives the device the very high input impedance, since the small-signal resistance between the gate and the channel is the same order of magnitude as a reverse-biased diode.

To understand the operation of the *n*-channel JFET, let us first consider the case where the gate and source are both at ground potential. In order to have the junction reverse biased, a positive potential is applied to the drain, as is shown in Fig. 7-3.

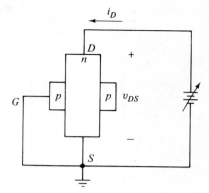

FIGURE 7-3 A circuit to obtain the *v-i* characteristics of a JFET with the gate at ground potential.

Initially, as the voltage v_{DS} is increased from zero to a small voltage, the channel acts as a resistor. The resulting value of resistance is called the channel resistance, which varies from unit to unit, ranging from about 500 Ω to about 2 kΩ in values.

As the voltage is increased from this small value to a larger value, the JFET acts as a reverse-biased diode, with the electrons in the *n*-type material tending to move away from the junction and all the holes in the *p*-type material moving toward the more negative region. (In this case, ground is more negative than the junction.) The result is the formation of a region void of carriers (holes and electrons), called *the depletion region*. This result is shown in Fig. 7-4a. Notice that the channel width W has decreased. Since the resistance of a material is inversely proportional to the cross-sectional area of the material, a decrease in W results in an increase in resistance. If the potential v_{DS} is still further increased, the width W goes to zero, as in Fig. 7-4b, and the resistance becomes very high. This potential is known as the *pinch-off voltage* V_p. Any increase in v_{DS} beyond V_p allows no additional drain current i_D to flow.

The resulting drain current that flows when the pinch-off voltage is reached is called I_{DSS} (the drain-to-source current with the gate shorted to the source). I_{DSS} is sometimes called *the saturation current* of the FET. The resulting voltage–current characteristics are shown in Fig. 7-5.

If the gate voltage is made negative, as shown in Fig. 7-6, the result is to make the depletion region larger for a given value of v_{DS}. Now it will take a smaller value of v_{DS} to reach the saturation current. The channel resistance is increased as well, since the channel width is narrower. The drain-to-source voltage at which the saturation will now be reached can be expressed as

$$v_{DS} \text{ (for saturation)} = V_p + v_{GS} \qquad (7\text{-}1)$$

Notice that the more negative v_{GS} is made, the smaller v_{DS} is when saturation first ocurrs. It is, of course, possible to make v_{GS} so large that the channel width is reduced to zero, even if no v_{DS} is applied. This occurs when v_{GS} is

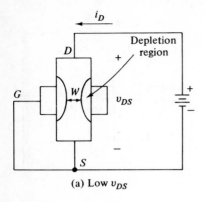

(a) Low v_{DS}

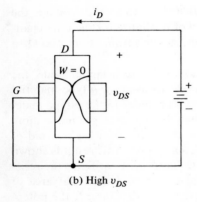

(b) High v_{DS}

FIGURE 7-4 The effect of increasing v_{DS} on the depletion region and the channel width W.

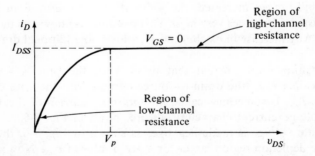

FIGURE 7-5 The $v_{DS} - i_D$ relationship of an FET with V_{GS} set at zero.

made equal to V_p. When this occurs, no drain current flows and we say that the FET is cut off. The typical characteristics in Fig. 7-7 show the v_{DS}–i_D relationship for various values of V_{GS}. These, of course, are the output characteristics of the FET.

Notice that for $V_{GS} = -2$ V in Fig. 7-7, the saturation of the drain cur-

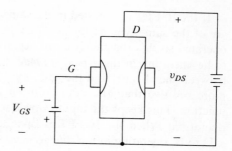

FIGURE 7-6 The JFET with a negative gate-source voltage.

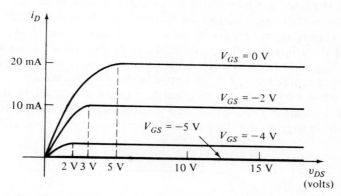

FIGURE 7-7 Typical output characteristics for an *n*-channel FET with $V_P = 5$ volts and $I_{DSS} = 20$ mA.

rent occurs at 3 V, since the pinch-off voltage is 5 V. At $V_{GS} = -5$ V, the drain current is approximately zero, since the width of the channel has been reduced to zero by the gate-to-source negative voltage.

The input characteristics for the JFET are obvious, since little or no gate current flows when the transistor is correctly biased. The input is essentially an open circuit.

The operation of the *p*-channel FET is the same as for the *n*-channel FET, except that all the voltage polarities are interchanged. The drain is connected to a negative supply, and the gate voltage must be more positive than the source.

7.2 Insulated Gate FET (IGFET)

The construction of the IGFET (or MOSFET) is similar to that of the JFET, except that a layer of silicon dioxide is placed between the *p*-type and the *n*-type materials. This results in increased input resistance and lower input capacitance.

If the IGFET is operated in the same way as the JFET, the output curves are of the same type as those of the JFET. When an *n* channel IGFET is operated so that the gate is more negative than the drain, the device is said to be operating in the *depletion mode*. However, with the layer of insulation between the channel and the gate, the IGFET can be operated with the gate more positive than the drain without making the input a forward-biased junction. This keeps the input impedance high and allows a larger region of operation. When the IGFET is operated in this manner, the transistor is said to be operating in the *enhancement mode*.

When the transistor is operated in the enhancement mode, the resulting field increases the conductivity of the carriers and allows more drain current to flow. Typical characteristics of an *n*-channel IGFET are shown in Fig. 7-8a, and the symbol for the device is shown in Fig. 7-8b. Notice that the value of v_{DS} at which saturation occurs has increased above the pinch-off voltage by the same amount as the applied gate-to-source voltage.

A word of *warning* should be given here to the newcomer to IGFET's. IGFET's are usually purchased with a shorting lead between the terminals to prevent damage to the silicon dioxide insulation from static charges. It

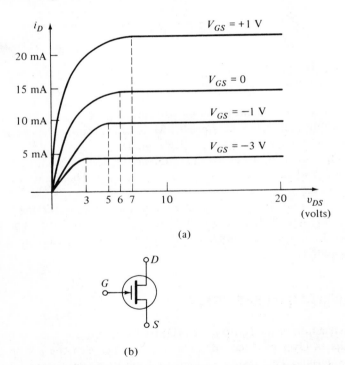

FIGURE 7-8 MOSFET output characteristics and symbol.

is well to do all the wiring or construction first, with the loop on the leads, and remove it *after* all the work has been completed. If the transistor must be handled without the loop, it should be by case only.

7.3 Equivalent Circuit for the FET

The equivalent circuit can be derived by considering the output current i_D to be a function of the other two variables, v_{DS} and v_{GS}, plotted in the output curves.

$$i_D = f(v_{DS}, v_{GS}) \tag{7-2}$$

The differential of the drain current is now found to be

$$di_D = \frac{\partial i_D}{\partial v_{DS}} dv_{DS} + \frac{\partial i_D}{\partial v_{GS}} dv_{GS} \tag{7-3}$$

and can be rewritten as

$$i_d = \frac{1}{r_d} v_{ds} + g_m v_{gs} \tag{7-4}$$

where it is understood that

$$i_d = di_D$$
$$v_{ds} = dv_{DS}$$
$$v_{gs} = dv_{GS}$$
$$\frac{1}{r_d} = \frac{\partial i_D}{\partial v_{DS}} \simeq \frac{\Delta i_D}{\Delta v_{DS}}\bigg|_{v_{GS}=\text{constant}}^{v_{gs}=0}$$

and

$$g_m = \frac{\partial i_D}{\partial v_{GS}} \simeq \frac{\Delta i_D}{\Delta v_{GS}}\bigg|_{v_{DS}=\text{constant}}^{v_{ds}=0}$$

In Eq. (7-4), r_d is called the *incremental drain resistance* and g_m is the *transconductance* of the FET.

Equation (7-4) can now be interpreted as a statement of Kirchhoff's current law and represented by the equivalent circuit shown to the right of the dotted line in Fig. 7-9. Figure 7-9 is the complete equivalent circuit of the FET. Notice that the gate-source terminals are left as an open circuit, since little or no current flows. A resistor r_g representing the gate-to-source resistance may be added, but for most applications this is very high and can be neglected.

The equivalent circuit shown in Fig. 7-9 is very appropriate, if the FET is operated in the saturation region (to the right of the pinch-off voltage) of the output characteristics. In this region, the curves are very flat and imply a practical current source that is dependent on the input voltage v_{gs}. The pa-

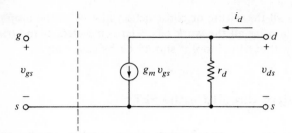

FIGURE 7-9 The equivalent circuit for an FET.

rameter g_m is the proportionality constant that relates the drain current to the applied voltage.

As in the case of h-parameters, the parameters g_m and r_d can be found from the output characteristics of the device. Consider the definition for g_m:

$$g_m = \frac{\partial i_D}{\partial v_{GS}} \simeq \frac{\Delta i_D}{\Delta v_{GS}}\bigg|_{\substack{v_{DS}=\text{constant} \\ =V_{DS}}} \tag{7-5}$$

Notice that g_m is the change in drain current divided by the change in gate-to-source voltage, keeping the drain-to-source voltage constant. Figure 7-10a

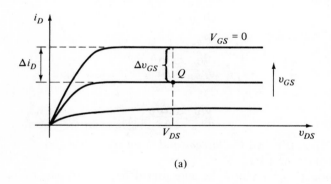

(a)

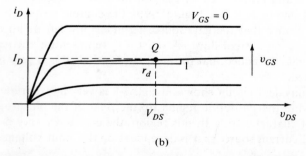

(b)

FIGURE 7-10 Obtaining g_m and r_d from the output characteristics.

indicates the method for obtaining g_m graphically. The incremental drain resistance is defined as

$$\frac{1}{r_d} = \frac{\partial i_D}{\partial v_{DS}} \simeq \frac{\Delta i_D}{\Delta v_{DS}}\bigg|_{\substack{v_{GS}=\text{constant} \\ =V_{GS}}} \tag{7-6}$$

Thus the parameter r_d is the change in drain-to-source voltage divided by the corresponding change in drain current for a constant value of v_{GS}. This parameter is the reciprocal of the slope in the output characteristics. Notice that, if the transistor is operated in the saturation region, r_d is a very large number. Typical values for r_d range from 50 kΩ to 1 MΩ. The reminder given here is that the small-signal parameters should be measured at the operating, or Q-point, of the JFET.

Typical values for the small-signal and other pertinent parameters for the JFET are given in Table 7-1.

TABLE 7-1 TYPICAL JFET PARAMETERS

Parameter	Value
I_{DSS}	10 mA
g_m	5 mA/V at $I_D = 8$ mA
r_d	80 kΩ
V_p	5 V

Example 7-1: Given a JFET having the characteristics listed in Table 7-1, sketch typical v–i output characteristics for this device.

Solution:

(a) The slope of the output characteristics is examined first. We assume that the characteristics will be plotted for all values of v_{DS} up to 20 V; then calculate the change in i_D that will occur for this range of values.

$$\Delta i_D = \frac{\Delta v_{DS}}{r_d} \\ = \frac{20 \text{ V}}{80 \text{ k}\Omega} = 0.25 \text{ mA} \tag{7-7}$$

The above value of Δi_D is quite small when compared to I_{DSS}, so that the lines will be assumed to be horizontal.

(b) The first line is now sketched with $V_p = 5$ V, $i_D = I_{DSS} = 10$ mA, and $V_{GS} = 0$ V, as shown in Fig. 7-11.

(c) Since g_m is equal to 5 mA V, the next line can be plotted. The drain current will now be approximately 5 mA less than I_{DSS} for V_{GS} equal to -1.0 V. The saturation of the drain current is reached at $v_{DS} = 4.0$ V. The resulting curve is added to Fig. 7-11.

(d) A final curve is added for v_{GS} equal to the pinch-off voltage of -5.0 V. The resulting drain current is zero. Additional curves may be sketched in

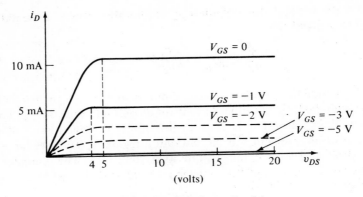

FIGURE 7-11 Solution to Ex. 7-1.

the graph, but these can only be estimated, since it is obvious that the value of g_m changes with i_D. These curves are shown as dotted lines in Fig. 7-11.

7.4 Biasing the FET

To bias the JFET we must remember that gate–source junction is to be reverse biased. In the case of *n*-channel FETs this means that the gate is to be negative with respect to the source, and the drain must operate with a positive voltage. A typical biasing circuit for the *common source amplifier* is shown in Fig. 7-12. This type of biasing is called *self-bias*, since it uses only one dc supply.

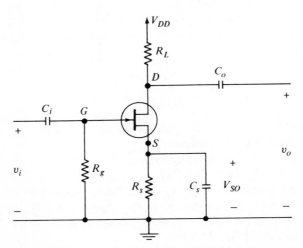

FIGURE 7-12 The biased common source amplifier.

The biasing and operation of FET may be likened to a BJT, if the gate is thought to be analogous to the base, the source to the emitter, and the drain to the collector. In Fig. 7-12, R_s provides a dc voltage V_{SO} to make the source more positive than the ground. Typically, this resistor is 50 Ω–1 kΩ. R_g shorts the gate to ground in a dc sense. Its typical value is 1 MΩ, a megohm being much smaller than the input impedance into the device. R_L has the usual function as a load resistor. Notice that since the gate is at ground potential, because of R_g, and the source is a positive potential V_{SO}, the gate is more negative than the source. To ensure that the source at ground potential is an ac sense, a large capacitor C_s is placed across R_s. C_i and C_o are the input and output coupling capacitors.

The biasing of FET can be best illustrated by means of a numerical example. Assume that we have a FET with the v–i characteristics shown in Fig. 7-13, and a supply voltage of 20 V. The first step is to decide on a suitable

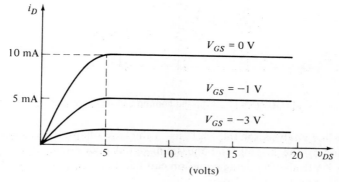

FIGURE 7-13 Typical n-channel JFET output characteristics.

drain current I_D so that we are in the operating region of the FET. $I_D = 5$ mA is such a value. To obtain this value, the voltage v_{GS} must be set at -1.0 V. Therefore, R_g is made 1 MΩ, and R_s is selected so that voltage drop across R_s is 1 V.

$$R_s = \frac{V_{GS}}{I_D} = \frac{1.0 \text{ V}}{5 \text{ mA}} = 200 \text{ Ω}$$

Now R_L is chosen so that voltage across R_L and V_{DS} are about equal. (This allows both positive and negative voltage variations on the drain.) Let us make the voltage across R_L 10 V. R_L is now calculated as

$$R_L = \frac{V_{RL}}{I_D} = \frac{10 \text{ V}}{5 \text{ mA}} = 2\text{k}Ω$$

This leaves 9 V between drain and source.

The capacitor C_s (across R_s) is chosen to be 50 μF. The capacitor C_i

can be much smaller, since it has to be a short circuit in comparison to the 1 MΩ at the input. A value of 0.1 or 0.01 μF is a typical choice for audio frequencies. If the output is to be fed into another FET amplifier, C_o can be about the same size as C_i. The resulting values of the components for a complete circuit are shown in Fig. 7-14.

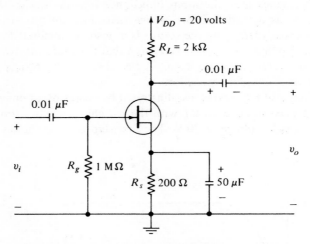

FIGURE 7-14 A biased common-source amplifier.

Load line analysis for the FET, as in the case of the BJT, is helpful for understanding the biasing and the small-signal operation of the amplifier. By using KVL, the following equation can be written for the dc operation of the circuit shown in Fig. 7-14:

$$V_{DD} = i_D(R_L + R_s) + v_{DS} \tag{7-8}$$

Solving for i_D we obtain

$$i_D = \frac{1}{R_L + R_s} v_{DG} + \frac{V_{DD}}{R_L + R_s} \tag{7-9}$$

Equation (7-9) is the equation of the dc load line for the FET amplifier and is plotted in Fig. 7-15 for the amplifier in Fig. 7-12. The method of plotting load lines for FETs is identical to that of the BJTs.

The ac load line is plotted in Fig. 7-15 as well. Notice that there is little difference in the slope between ac and dc load lines. The reason for this is that R_s is small in comparison to R_L, and, since the slope of the ac load line is $1/R_L$, as compared to $1/(R_L + R_s)$ for the dc load line, they are essentially the same.

It is interesting to note that once I_D and V_{GS} have been chosen, the bias voltage V_{DS} depends only on the choice of R_L. Selecting R_L larger or smaller

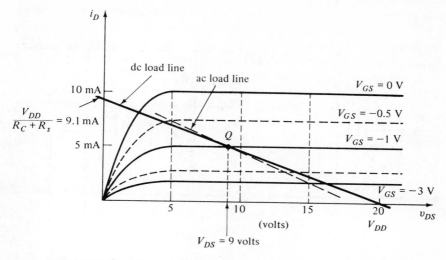

FIGURE 7-15 A numerical example of an ac and dc load line for a common source amplifier.

has the effect of decreasing or increasing V_{DS} only. The reason for this is the flatness of the output curve, which implies a practical current source. Since the output is a practical current source, the voltage across R_L is determined solely by the value of R_L.

It is also interesting to note that the input sinusoidal signal should be small for the output to be sinusoidal. The reason for this is that the lines in output curves are not as evenly spaced as they are for the BJT. There is, nevertheless, one simpler property of FET amplifier and that is the fact that the input current need not be calculated. As soon as the amplitude of the input voltage is known, the output voltage can be determined directly from the ac load line by traversing along the ac load line, up and down from the Q-point.

Biasing the IGFET can be done in the same fashion as the JFET, if the transistor is used in the depletion mode. If the IGFET operates in both the depletion and enhancement modes, it can be made to operate at *zero bias*, with the grid-to-source voltage set at zero V. This can be accomplished very simply by the circuit shown in Fig. 7-16a, the design procedure being the same as in the example discussed previously.

If the IGFET is to be biased in the enhancement mode, this can be accomplished by the circuit shown in Fig. 7-16b. Here the design procedure is the same as that for H-type biasing of a BJT, except that a large resistance, R_g, is placed between the gate and point X, as shown in Fig. 7-16b, to raise the input impedance to an acceptable level. This can be done, since no gate current flows.

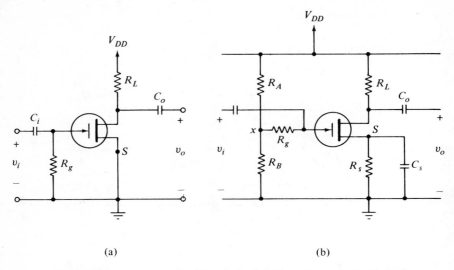

FIGURE 7-16 Biased IGFET in the common source configuration. (a) zero bias (b) bias in the enhancement mode.

7.5 Small-Signal Analysis of an FET Amplifier

As in the case of BJT amplifiers, the ac equivalent circuit derived in Section 7.4 is used to calculate the input impedance, output impedance, and voltage gain. The most important property is that of voltage gain, since, in cascading FET amplifiers, loading is not a severe problem because of the high input impedance.

Consider the common source amplifiers shown in Fig. 7-17a. If the usual approximations are made in assuming the capacitors to be large, and short circuits to ac, the equivalent circuit shown in Fig. 7-17b is obtained. The resistance r_d is neglected since it is usually much larger than R_L.

The output voltage v_o becomes

$$v_o = -g_m v_{gs} R_L \tag{7-10}$$

Since $v_{gs} = v_i$, the voltage gain can be calculated from Eq. (7-10), as

$$A_v = \frac{v_o}{v_{gs}} = \frac{v_o}{v_i} = -g_m R_L \tag{7-11}$$

If typical values for $g_m R_L$ are substituted in Eq. (7-11), we find that the range of values of $|A_v|$ is between 3.0 and 15.0 approximately. This is quite a bit lower than the gain achieved for a CE amplifier; however, the input impedance is much higher as found below.

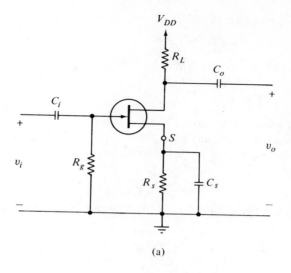

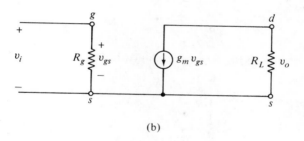

(a) The common source amplifier and
(b) its equivalent circuit.

FIGURE 7-17 FET common source amplifier and equivalent circuit.

The input impedance is, of course, simply equal to R_g.

$$R_{in} = R_g \tag{7-12}$$

Typically, R_{in} is selected between 1 and 10 MΩ for a JFET, and much higher for an IGFET.*

The output impedance, found in the usual fashion, is given by

$$R_{out} = R_L \tag{7-13}$$

*The maximum value of R_g to the used with an IGFET is usually specified by the manufacturer.

Example 7-2: Given a *p*-channel JFET with the following circuit parameters:

$$I_{DSS} = -20 \text{ mA}$$
$$g_m = 10 \text{ mA/V at } I_D = -10 \text{ mA} \quad (7\text{-}14)$$
$$r_d = 100 \text{ k}\Omega$$
$$V_p = -5 \text{ V}$$

In addition, a power supply of 20 V is available.
(a) Design a common source amplifier with a voltage gain of -10.
(b) Find the input and output impedance of the amplifier designed in part (a).

Solution:

(a) To obtain the required gain, the value of R_L must be determined first:

$$R_L = \frac{-A_v}{g_m} = \frac{10}{10 \text{ mA/V}} = 1 \text{ k}\Omega \quad (7\text{-}15)$$

To obtain a reasonable Q-point, the FET can now be biased at $I_D = -10$ mA. To achieve this current, v_{GS} should be set at $+1.0$ V. Hence

$$R_s = \frac{V_{so}}{I_D} = \frac{-1 \text{ V}}{-10 \text{ mA}} = 100 \text{ k}\Omega \quad (7\text{-}16)$$

We now have the circuit, as shown in Fig. 7-18.

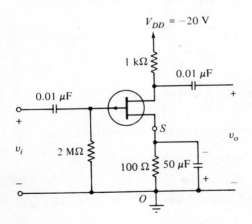

FIGURE 7-18 Circuit for the solution to Ex. 7-2.

(b) The input impedance is equal to R_g. In this case, 2 MΩ was chosen.

$$R_{in} = R_g = 2 \text{M}\Omega \quad (7\text{-}17)$$

The output impedance is given by

$$R_{out} = R_L = 1 \text{ k}\Omega \quad (7\text{-}18)$$

7.6 Common Drain Amplifier (Source Follower)

The source follower is analogous to the emitter follower. It offers a very high input impedance, unity gain, and a relatively low output impedance. The major difference between the source follower and the emitter follower is that it need not be biased by a previous stage to exhibit the high input impedance.

The JFET is usually the device that is used in making a source follower, since an IGFET, common source amplifier, already has a very high input impedance. IGFET's are generally more expensive, and, in many cases, it may be cheaper to use an additional JFET source follower to obtain a higher input impedance.

Figure 7-19 illustrates a circuit that could be used to obtain a practical

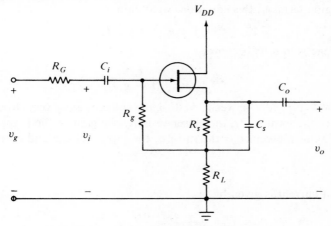

FIGURE 7-19 AN FET source follower circuit.

source follower. The proper gate–source bias voltage is supplied by the resistor R_s. The biasing procedure is the same as for the common source amplifier. R_G and v_g represent the Thévenin equivalent of source driving the emitter follower. v_i is the applied voltage to the source follower.

If the usual assumptions are made, the circuit shown in Fig. 7-20 is the equivalent circuit for the source follower. Using KVL, we can now write

$$v_i = v_{gs} + v_o \qquad (7\text{-}19)$$

or

$$v_{gs} = v_i - v_o \qquad (7\text{-}20)$$

Since $R_g \gg R_L$, most of the current from the current generator $g_m v_{gs}$ will flow through R_L, so that the output voltage v_o becomes

$$v_o = g_m v_{gs} R_L \qquad (7\text{-}21)$$

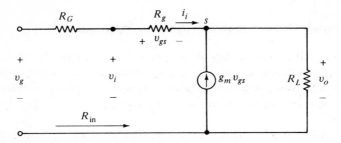

FIGURE 7-20 The equivalent circuit for the FET source follower.

Substituting Eq. (7-20) into Eq. (7-21), we arrive at

$$v_o = g_m(v_i - v_o)R_L \tag{7-22}$$

Rearranging terms in this equation, we obtain

$$v_o(1 + g_m R_L) = g_m v_i R_L \tag{7-23}$$

The voltage gain v_o/v_i becomes

$$A_v = \frac{v_o}{v_i} = \frac{g_m R_L}{1 + g_m R_L} \tag{7-24}$$

Since $g_m R_L$ is typically between 2 and 15, the voltage gain is very close to unity. The larger the term $g_m R_L$, the closer to unity the gain of the amplifier.

The input impedance is the ratio of the input voltage and current:

$$R_{in} = \frac{v_i}{i_i} \tag{7-25}$$

The input current is given by

$$i_i = \frac{v_i - v_o}{R_g} = \frac{v_i(1 - A_v)}{R_g} \tag{7-26}$$

Substituting Eq. (7-26) into Eq. (7-25), we obtain

$$R_{in} = \frac{v_i}{i_i} = \frac{R_g}{1 - A_v} \tag{7-27}$$
$$= R_g(1 + g_m R_L)$$

Since the term $(1 + g_m R_L)$ in Eq. (7-27) is always greater than unity, it follows that the input impedance may be several times greater than R_g.

The output impedance can be found by finding the ratio of open-circuit voltage and short circuit current:

$$R_{out} = \frac{v_{oc}}{i_{sc}}$$

The open-circuit voltage is simply

$$v_{oc} = A_v v_i \tag{7-28}$$

The short-circuit current is given by

$$i_{sc} = g_m v_{gs} = g_m v_i \tag{7-29}$$

It is interesting to note that in Eq. (7-29) the short-circuiting of the output in Fig. 7-20 makes the grid-to-source voltage equal to the input voltage v_i. The ratio of Eqs. (7-28) and (7-29) produces the expression for the output impedance:

$$R_{out} = \frac{A_v}{g_m} = \frac{R_L}{1 + g_m R_L} \tag{7-30}$$

Upon examination of Eq. (7-30), it is interesting to note that output impedance of the source follower is independent of the source impedance. For a given transistor (i.e., a given g_m), the output impedance is dependent solely on the choice of R_L. The smaller R_L, the smaller the output impedance. This result is certainly quite different from the type of result obtained in the emitter follower.

The source follower of Fig. 7-19 is not the most commonly used form. It is usually common to find that the capacitor C_s is omitted, since R_s is much smaller than R_L and is practically a short circuit to ac. This is, of course, an approximation, but the formulas for input impedance, output impedance, and voltage gain are usually assumed to be the same as the ones derived previously in this section. It is left as an exercise to derive more accurate equations for the properties of the source follower, which include the effect of the resistor R_s. The more common form of the source follower is shown in Fig. 7-21.

As in the case of the BJT emitter follower, the source follower can be directly coupled to a previous stage. Two examples of this type of interconnection are shown in Fig. 7-22. In these cases, biasing is even easier than for the self-biased source follower. Since the voltage $|V_{GS}|$ is usually much

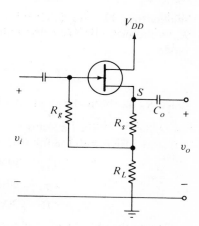

FIGURE 7-21 A practical FET source follower circuit.

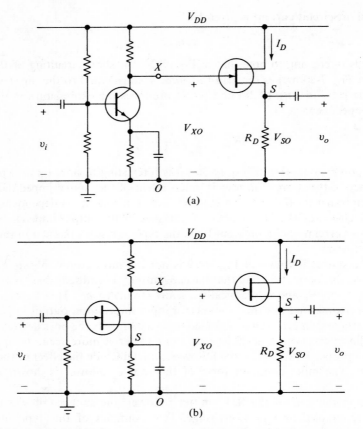

FIGURE 7-22 Direct coupled FET source followers: (a) from a bipolar transistor amplifier (b) from another FET amplifier.

smaller than V_{XO}, it is assumed that $V_{XO} \simeq V_{SO}$. To bias the source follower, R_D is chosen so that I_D is less than I_{DSS}, to enable operation in the depletion mode:

$$R_D \simeq \frac{V_{XO}}{I_D} \quad \text{where } I_D < I_{DSS} \quad (7\text{-}31)$$

Once R_D has been chosen, V_{GS} adjusts itself so that the proper I_D flows.

Example 7-3: Given a source follower shown in Fig. 7-23, biased with an *n*-channel FET with the following parameters:

$$I_{DSS} = 10 \text{ mA}$$
$$g_m = 5 \text{ mA/V at } I_D = 5 \text{ mA}$$
$$V_p = 3 \text{ V}$$

Calculate the voltage gain, input impedance, and output impedance.

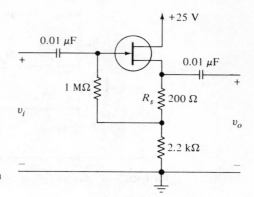

FIGURE 7-23 Circuit for the solution of example problem 7-3.

Solution:

Even though R_s is unbypassed, we shall assume the effect of R_s to be negligible and use formulas previously derived.

(a) Voltage gain:

$$A_v \simeq \frac{g_m R_L}{g_m R_L + 1} = \frac{5 \times 10^{-3} \times 2.2 \times 10^3}{5 \times 10^{-3} \times 2.2 \times 10^3 + 1}$$
$$\simeq \tfrac{11}{12} = 0.92$$

(b) Input impedance:

$$Z_{in} \simeq R_g(1 + g_m R_L) = 12 R_g = 12 \text{ M}\Omega$$

(c) Output impedance:

$$Z_{out} \simeq \frac{R_L}{1 + g_m R_L} = \frac{R_L}{12} = \frac{2.2 \text{ k}\Omega}{12}$$
$$= 185 \text{ }\Omega$$

7.7 Common Gate Amplifier

The common gate amplifier is not frequently used at audio frequencies. Its use is for amplifiers that have to operate over a large range of frequencies and is generally used for high-frequency RF amplifiers.

The common gate amplifier is analogous to the common base amplifier. Biasing the common gate amplifier is the same as for the common drain amplifier. A typical common gate amplifier is shown in Fig. 7-24. The input impedance, output impedance, and voltage gain can be found easily by the use of the equivalent circuit for an FET. These properties are

$$R_{in} = \frac{R_s}{1 + g_m R_s} \qquad (7\text{-}32)$$

$$R_{out} = R_L \qquad (7\text{-}33)$$

$$A_v = +g_m R_L \qquad (7\text{-}34)$$

176 Chap. 7 / The Field Effect Transistor

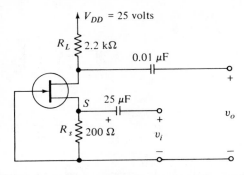

FIGURE 7-24 An example of a common gate amplifier.

The input impedance is relatively low, the voltage gain is the same as for the common drain amplifier, except for the opposite sign, and the output impedance is equal to the load resistor.

7.8 FET as a Voltage Variable Resistor

For all the FET amplifier circuits considered previously, the FET was biased in the "flat region" of the output v–i curves. This region is shown as the shaded area in Fig. 7-25 and is usually known as the *pentode operating region*. The FET can also be used in the region to the left of the ($i_D = V_p + v_{GS}$) curve. This region is usually called the *triode operating region* of the FET.

If the FET is biased at the origin (i.e., $V_{DS} = 0$), it exhibits the same properties in the third quadrant as in the first, since the source and drain can usually be interchanged. For a given value of $v_{GS} = V_{GS}$, a resistance r_c (the

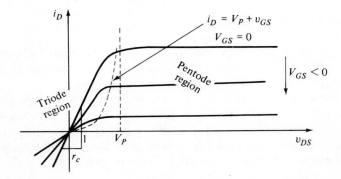

FIGURE 7-25 The output characteristics of an FET shown as two separate regions: the triode region and the pentode region.

channel resistance) can be defined:

$$r_c = \frac{\Delta v_{DS}}{\Delta i_D}\bigg|_{\substack{v_{GS}=V_{GS} \\ V_{DS}=0}} \quad (7\text{-}35)$$

r_c is the reciprocal of the output curves at origin and can be varied by varying v_{GS}. A typical value for r_c at v_{GS} set at zero is usually 500 Ω. If v_{GS} is made more negative by several volts, r_c will change by a factor of 1 or 2. The result is a voltage-controlled variable resistor.

A device such as this finds many applications in electronics. It can be used as a modulator, an automatic volume control, or in a tremolo circuit, to name a few.

An interesting application of the FET as a voltage-controlled resistor is shown in Fig. 7-26 as a tremolo for an electronic organ. A tremolo is a device

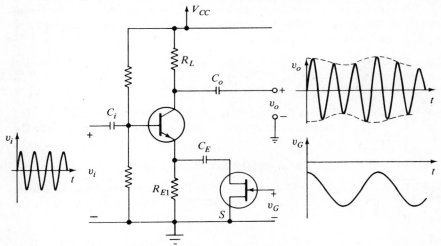

FIGURE 7-26 An amplifier with variable gain controlled by a voltage.

that varies the output amplitude sinusoidally at the rate of a few cycles per second to produce an interesting musical effect.

The tremolo effect is achieved by varying the gain of the amplifier with the voltage v_G. Since the gain of the amplifier is given by

$$A_v = \frac{v_o}{v_i} = -\frac{R_L}{R_E} \quad \text{where } R_E = R_{E_1} \| r_c \quad (7\text{-}36)$$

and since r_c is a function of v_G (the gate-to-source voltage of the FET), then R_E is a function of v_G and hence the gain can be written as

$$A_v = \frac{-R_L}{R_E(v_G)} \quad (7\text{-}37)$$

Hence the amplifier has a variable gain that is a function of the voltage v_G. If an *n*-channel JFET is used, v_G should be a sinusoid biased in the negative direction. On the other hand, the use of an MOSFET at zero bias would allow v_G to be a pure sinusoid without a bias voltage.

PROBLEMS

7.1 Sketch typical output characteristics for an *n*-channel JFET with the following parameters: $V_p = 3$ V, $I_{DSS} = 6$ mA, $g_m = 3$ mA/V at $I_D = 5$ mA, and $r_d = 50$ kΩ.

7.2 (a) Bias the FET in Problem 7.1 to obtain a voltage gain of 12, using the common source configuration.
(b) Add a source follower to the amplifier in part (a), using the same transistor. Use a 3-kΩ load resistor in the source follower. Calculate the resulting output impedance.

7.3 Sketch typical output characteristics for an *n*-channel MOSFET with the following parameters: $V_p = 3$ V, $I_{DSS} = 6$ mA, $g_m = 3$ mA/V at $I_D = I_{DSS}$, and $r_d = 100$ kΩ.

7.4 Bias the MOSFET in Problem 7.1 in the enhancement mode at $V_{GS} = +0.5$ V. Use the transistor in the common source configuration and estimate the gain that will be obtained.

7.5 Use a FET and BJT to design the amplifier in Problem 6.15.

7.6 Design an amplifier with a voltage gain of 100, an input impedance of 1 MΩ, and an output impedance of less than 5 kΩ. (*Hint:* More than one stage may be used.)

7.7 Derive the equations for R_{in}, R_{out}, and A_v for the common gate amplifier.

7.8 Design a common gate amplifier to have an input impedance of 200 Ω and a voltage gain of 8.

7.9 Design a difference amplifier such as the one shown in Fig. P7-9, using the transistor specifications of Problem 7.1.

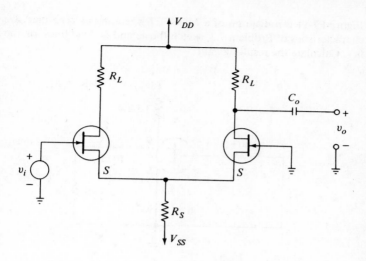

FIGURE P7-9

7.10 Analyze the circuit shown in Fig. P7-10.

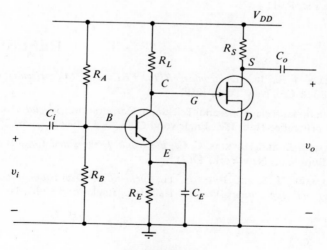

FIGURE P7-10

180   Chap. 7 / The Field Effect Transistor

7.11 Figure P7-11 is a diagram of a MOSFET operating at zero bias. Using the characteristics of Problem 7.3, sketch the ac and dc load lines for the amplifier. Calculate the resulting gain.

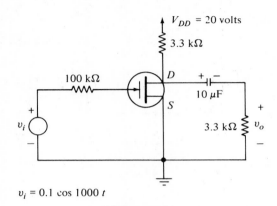

$v_i = 0.1 \cos 1000\, t$

FIGURE P7-11

7.12 Why is a coupling capacitor unnecessary for a FET amplifier if it is being driven by sinusoidal source with an output impedance of 1 MΩ or less? (See Fig. P7-11.)

REFERENCES

[1] ANGELO, J. E., JR., *Electronics: BJTs, FETs, and Microcircuits* (McGraw-Hill Book Co., New York, 1969).

[2] International Telephone and Telegraph Corporation, *Special Purpose Transistors* (Prentice-Hall, Inc., Englewood Cliffs, N.J., 1966).

[3] MILLMAN, J., and HALKIAS, C. C., *Electronic Devices and Circuits* (McGraw-Hill Book Co., New York, 1967).

[4] WALLMARK, J. T. and JOHNSON, H., *Field-Effect Transistors, Physics, Technology and Applications* (Prentice-Hall, Inc., Englewood Cliffs, N.J., 1966).

8

AIDS FOR DESIGN

The design of electronic circuits is indeed an interesting field of endeavor. Seldom are two design problems the same. Because requirements for individual circuits vary so much, it is difficult to teach design. At best, a text on design can only list a few of the types of problems that are likely to be encountered. After studying these, the student designer must rely on his own ingenuity to solve other problems.

A good circuit designer usually keeps a "mental dictionary" of the basic and common circuits so that he can recall them quickly when the need arises. The other more complicated and novel circuits are usually kept in a notebook for use in the more complicated problems.

An attempt has been made in the first seven chapters to supply the reader with such a dictionary of the fundamental circuits, as well as with the methods for using them successfully.

8.1 Circuit Specifications

Circuit specifications are required information that must be obtained before a circuit can be designed. The designer usually has little or no control over circuit specifications, since these arise from the circuit function in a larger system.

For example, consider a problem that is typically encountered. An engineer has purchased an audio amplifier with speakers and a microphone. Upon connecting the amplifier to the microphone and speakers, he discovers that the output signal from the microphone is not large enough to drive the amplifier and speakers sufficiently. The engineer must design a preamplifier to make the system operate properly. Before he can do so, he must know the output impedance of the microphone, the input impedance of the audio amplifier, and the voltage gain needed to operate the audio amplifier properly. He then translates this information into a specification list that probably contains most of the following information:

1. The desired gain.
2. The input and output impedance.
3. The load and source impedance.
4. The signal levels at the input and output of the circuit.
5. The phase-shift, or transient, response.
6. The voltage(s) of the power supply(s).
7. The current drain from the supply(s).
8. The noise and distortion level that is acceptable.
9. The sensitivity of the circuit to parameter changes or component replacement.
10. The temperature-operating range.
11. The size, weight, and appearance of the package.
12. The durability of the package (environmental specifications such as shock, moisture, dust, vibration, etc.).
13. The effect of the product on the human operator (safety, ease of operation, physical strain, etc.).
14. The cost.

The above list is a typical set of specifications that must be met. In any particular design problem, most of the above items have to be considered, but some may become more important than others. In the above example, the first four items are of primary importance. Items 11 and 12 may become very important as well if the amplifier is to be used by many different people.

Many design examples can be constructed by specifying any or all of the above items. The remainder of the text contains worked examples in which different items are stressed as new approaches are developed. In addition, the student can help himself by studying recent journals, which are filled with different design approaches.

It is well worth mentioning that in many, if not in most, practical problems, it is difficult to achieve *all* the circuit specifications. In these cases, the designer must exercise his judgment in arriving at compromised solutions by meeting the more important specifications first.

8.2 Transistor Specifications

During the process of design, in trying to meet circuit specifications, the designer will probably have to choose the type of transistor he will use. There are virtually thousands from which to choose, and most manufacturers produce a large variety of different types. The choice will depend on the role of the transistor needed in the circuit.

The type of transistor finally chosen will depend on its specifications. The following list of parameters are typical of the ones that can be obtained from a manufacturer on request.

1. *Beta* (β) or h_{FE}

    Transistors come in low, high, and medium values of beta, ranging from about 40 to about 300. A value of 100 is considered to be a medium value.

2. *Maximum Collector Power Dissipation* P_d

    This value is usually specified at 25°C (room temperature). It is generally from about 60 to 250 mW for the small-signal transistor but may be as high as several hundred watts for larger power transistors.

3. *Maximum Collector Breakdown Voltages* BV_{CEO}

    There are several types of breakdown phenomena that can occur in a transistor in the common emitter configuration. These are usually given as BV_{CEO}, BV_{CEX}, and BV_{CER}. The quantities are quite simple to understand after the notation has been defined: The BV designates a *breakdown voltage*; the first two subscripts indicate the two points at which the voltage is applied, with the second subscripting symbol indicating the common terminal; the third subscript indicates the manner in which the third terminal is connected. If the third subscript is O, then the third terminal is an open circuit; if the third subscript is X or S, then the terminal is short-circuited to the common terminal; and if the subscript is R, then it means that a resistor is connected between the third terminal and the common point. The value of this resistor must be specified. Thus the symbol BV_{CEO} is the maximum collector-to-emitter voltage with the base left open-circuited.

    BV_{CEO} is the value that is usually specified by the manufacturer. This quantity ranges from as low as 6 V for some transistors to as high as 600 V for others.

    Figure 8-1a illustrates the different types of collector-to-emitter breakdown that can occur. BV_{CEX} is usually 10 to 30 per cent higher than BV_{CEO}, with BV_{CER} falling between those two values, depending on the value of R that is used.

    Figure 8-1b is a photograph of a typical transistor output characteristic taken from a transistor curve tracer. The breakdown voltage BV_{CEO} is clearly shown and is about 55 V.

    For designing amplifiers it is wise to choose transistors with BV_{CEO} at least 20 per cent higher than the largest value of V_{CC} to be used. This almost ensures that the Q-point is out of the *breakdown region* shown in Fig. 8-2.

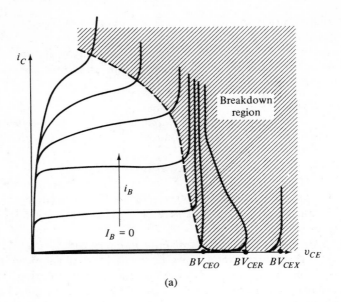

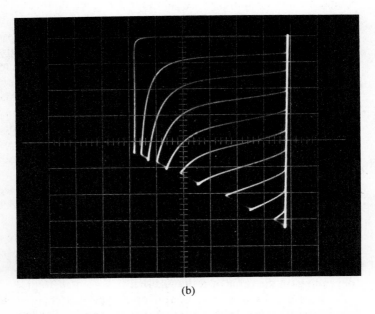

FIGURE 8-1 (a) Collector characteristics illustrating breakdown voltages BV_{VEO}, BV_{CER}, BV_{CEX}. (b) Typical transistor output characteristics. (Scales: Vertical $(i_C) = 1$ mA/div; horizontal $(v_{CE}) = 10.0$ V/div; and running parameter $(i_B) = .005$ mA/step.)

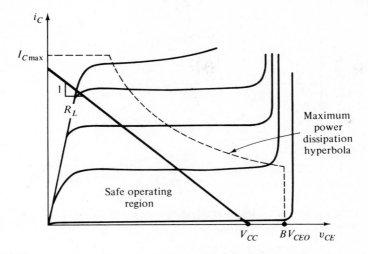

FIGURE 8-2 Collector characteristics illustrating safe operating region for transistor when I_{Cmax}, BV_{CEO}, and the maximum power dissipation hyperbols are specified.

4. *Maximum Collector Current I_{Cmax}*

The maximum collector current is self-explanatory, and, together with BV_{CEO} and P_d, I_{Cmax} completes the safe operating region of the common emitter collector characteristics. For small-signal amplifiers, the location of the load line and Q-point in this region ensures safe operation. Such a well-chosen load line is shown in Fig. 8-2.

5. *Gain–Bandwidth Product f_T*

In all our analysis thus far, we have assumed the small-signal parameters to be constant with frequency. This is approximately true for the audio frequencies but at higher frequencies the parameters undergo change. The current-gain β, for example, decreases with increasing frequency. A typical log–log plot of β as a function of frequency is shown in Fig. 8-3. The frequency at which β degenerates to the value 1.0 is known as f_T, or the gain–bandwidth product of a transistor.

The frequency of f_T is usually specified by the manufacturer. Typical values of f_T range from about 1 or 2 megahertz to about 1 or 2 gigahertz, depending on the transistor type.

6. *Maximum Temperature*

The maximum temperature at which a transistor can be operated without destruction of the junctions is specified by the manufacturer. This temperature is usually between 70 and 140°C.

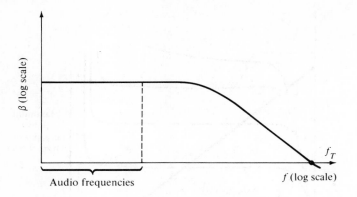

FIGURE 8-3 A sketch illustrating the variation of magnitude with frequency.

7. *Size and Mounting Case*
8. *Cost*

The cost factor cannot be overemphasized, since price variations of transistors are so great. A designer must often choose between the cost of the transistor and the type of specifications to be used.

9. *Leakage I_{CBO} and I_{CEO}*

In Eq. (5-10) the leakage factor I_{CBO} was defined as the amount of collector current flowing when the emitter current was set equal to zero, that is,

$$i_C = \alpha i_E + I_{CBO} \tag{5-10}$$

The emitter current as in Fig. 5-20 is given by

$$i_E = i_C + i_B \tag{8-1}$$

and, substituting Eq. (5-10) into Eq. (8-1), we obtain

$$i_E(1 - \alpha) = i_B + I_{CBO} \tag{8-2}$$

Solving for i_E we obtain

$$i_E = \frac{1}{1 - \alpha} i_B + \frac{I_{CBO}}{1 - \alpha} \tag{8-3}$$

Since $i_E = i_B + i_C$, Eq. (8-3) becomes

$$i_C = i_B \left(\frac{1}{1 - \alpha} - 1 \right) + \frac{I_{CBO}}{1 - \alpha} \tag{8-4}$$

or

$$i_C = \beta i_B + (\beta + 1) I_{CBO} \tag{8-5}$$

since

$$\frac{1}{1 - \alpha} = \beta + 1 \tag{8-6}$$

Equation (8-5) is usually rewritten as

$$i_C = \beta i_B + I_{CEO}$$

where $I_{CEO} = (\beta + 1)I_{CBO} = (h_{FE} + 1)I_{CBO}$

I_{CEO} is the collector leakage current that flows in the common emitter configuration when the base is open-circuited. Notice that this current is $\beta + 1$ times larger than I_{CBO}. Figure 8-4 shows the

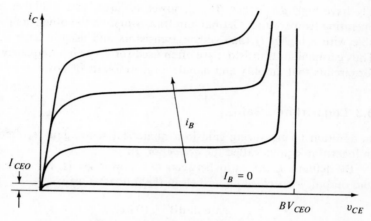

FIGURE 8-4 Collector characteristics illustrating I_{CEO}.

method for obtaining I_{CEO} from the manufacturer's CE output characteristics or the curve tracer display for a given transistor. In most small-signal transistors ($P_d < 1$ W), I_{CEO} is much less than 1 μA and is usually neglected in most design procedures. However, in large power transistors, I_{CEO} may be as large as several milliamperes and it may be necessary to consider this leakage current in the design procedure.

The parameter I_{CEO} is usually specified by the manufacturer, although I_{CBO} is sometimes specified. These two parameters are generally referred to as the common emitter, or common base, leakage currents, respectively.

10. *Noise*

We have not previously dealt with the problem of noise in circuits, and this is a good time to mention it. For many transistors (to be used in low-noise circuits) the noise figure (*NF*) is given at a specified frequency, bias point, and source resistance.

11. *h-Parameters*

Many specifications give typical *h*-parameters at a specified *Q*-point. These are helpful, although not essential for design work.

12. *High-Frequency Parameters*

For transistors to be used at frequencies above the audio range, other small-signal parameters are given. These are usually plotted on graphs as functions of bias voltage and current as well as functions of frequency. These quantities will be examined in more detail in later chapters.

One of the choices that usually has to be made in transistor circuit design is whether to use a germanium or silicon transistor. Silicon transistors generally have high βs, higher BV_{CEO}, lower leakage, and a higher maximum operating temperature. Germanium transistors, on the other hand, are available with a higher f_T than silicon transistors and usually have a larger P_d. Thus germanium transistors are often used for very high frequency work and for circuits that amplify and handle large amounts of power.

8.3 Logarithmic Gain

In addition to measuring gain in a numerical sense, gain is often specified in logarithmic units called the *decibel* or dB.

By definition, the ratio between two power levels, for example, input and output, is specified in decibels in the following way:

$$\text{gain in dB} = 10 \log_{10} \frac{P_o}{P_i} \quad (8\text{-}7)$$

where P_o represents the output power and P_i represents the input power into a device.

If the input and output impedances are resistive and equal to R_{in} and R_{out}, then the gain in dB in Eq. (8-7) can be expressed as the ratio of input and output voltages in the following way:

$$\text{gain in dB} = 10 \log_{10} \frac{\frac{V_o^2}{R_{out}}}{\frac{V_i^2}{R_{in}}} \quad (8\text{-}8)$$

where V_o is the output voltage and V_i is the input voltage.

If $R_{in} = R_{out}$, Eq. (8-8) becomes

$$\text{gain in dB} = 20 \log_{10} \frac{V_o}{V_i} \quad (8\text{-}9)$$

Equation (8-9) is the power gain in dB provided the input and output impedances are equal. However, for ordinary transistors, this seldom is the case, so that this definition has been accepted as the voltage gain in dB. A more precise definition can be used, namely redefining a new unit of dBv (decibels of voltage):

$$\text{gain in dBv} = 20 \log_{10} \frac{V_o}{V_i} \quad (8\text{-}10)$$

This term is seldom used in practice and simply referred to as dB.

There are many reasons for using dB rather than numerical gain. One is that the human senses have almost a logarithmic response, so that dB units are more appealing to operators of electronic equipment. For example, an audio amplifier driving a loudspeaker and producing twice as many dB as another would seem to be twice as loud as the first. Another reason for using dB is that the gain of a cascade of nonloading stages is the sum in dB units rather than the product, making the overall calculation easier.

8.4 Use of Frequency-Response Plots

The following two sections are not intended as a comprehensive study of frequency-response plots, but rather a review of the essentials necessary for use in electronic design. A knowledge of the method of solving simple differential equations as applied to simple electrical networks by the operator or Laplace transform method is necessary. Readers totally unfamiliar with these concepts should read some of the material suggested in the references at the end of this chapter.

Consider the two-port network shown in Fig. 8-5. All the terminal voltages and currents are designated in uppercase letters. This is because we are

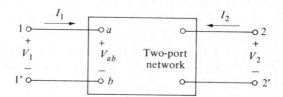

FIGURE 8-5 Voltage and current reference directions for a typical two port network.

considering the *transforms* of the actual quantities. In electronics it is common to use lettered subscripts rather than numerical ones, as is done in the discipline of circuit theory. Thus a voltage such as V_1 would be designated as V_{ab} in accordance with the symbolism rules already established, as shown in Fig. 8-5.

For the two-port shown in Fig. 8-5, a *transfer function* is any ratio of the port variables. These have mathematical properties similar to *impedance functions*. The network functions that we are most concerned with in electronics are the voltage gain, input impedance, and output impedance. In general, the network functions are functions of frequency and take on different values as the frequency of the input signal is varied. The method of Bode plots enables us to sketch the approximate value of these functions as a function of frequency. Bode plots are usually plotted on log–log paper with frequency as one variable and the amplitude or phase as another variable.

Consider the network shown in Fig. 8-6a. Let us examine the voltage

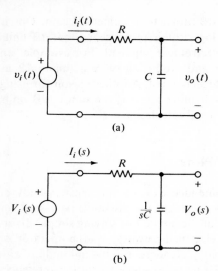

(a) An RC network in the time domain
(b) The same RC network in the complex frequency domain

FIGURE 8-6 RC network in the time domain and the complex frequency domain.

gain A_v as a function of frequency. The first step is to redraw Fig. 8-6a with all the voltages, currents, and impedances in the complex frequency domain, as shown in Fig. 8-6b. Notice that the voltages and currents are a function of the complex frequency variable s,* which becomes $j\omega$ in the steady state. The impedance of the capacitor becomes $1/sC$. The voltage gain can now be calculated by voltage division:

$$A_v(s) = \frac{V_o}{V_i} = \frac{\frac{1}{sC}}{R + \frac{1}{sC}} \quad (8\text{-}11)$$

$$= \frac{1}{1 + sRC}$$

If phasor analysis is used for sinusoidal input voltages, Eq. (8-11) becomes

$$A_v(j\omega) = \frac{1}{1 + j\omega\tau} \quad (8\text{-}12)$$

where

$$\tau = RC$$

The above expression is easily obtained by substituting s by $j\omega$ in Eq. (8-11).

*The variable s here is the Laplace transform variable. If the reader is not familiar with these concepts, s can be viewed as being the generalized form of the variable $j\omega$ and is the same if only the steady-state solution is required, as in phasor analysis.

Notice that Eq. (8-12) is now a function of the real frequency variable ω.

To gain an understanding of how the voltage gain varies with frequency, we observe the circuit action first at very low frequencies and then again at very high frequencies. At low frequencies when ω is near zero, or $\omega\tau \ll 1$, the expression for voltage gain in Eq. (8-12) can be approximated to be

$$A_v \simeq 1 \qquad \text{for } \omega\tau \ll 1 \text{ or } \omega \ll \frac{1}{\tau} \qquad (8\text{-}13)$$

At high frequencies, when ω is very large, or $\omega\tau \gg 1$, the voltage gain is approximated as

$$A_v(j\omega) \simeq \frac{1}{j\omega\tau} \qquad \text{for } \omega\tau \gg 1 \text{ or } \omega \gg \frac{1}{\tau} \qquad (8\text{-}14)$$

Equation (8-14) indicates a decrease in the magnitude of the gain with increasing frequency. The phase is approximately 90° for large ω.

The results of Eqs. (8-13) and (8-14) can be observed to be reasonable results from a nonmathematical point of view. At low frequencies the reactance of the capacitor is very large (practically an open circuit), and the output voltage becomes equal to the input voltage. At high frequencies the reactance of the capacitor decreases with increasing frequency, and the resulting gain decreases as well.

To observe the changing gain with frequency graphically, Eq. (8-12) is plotted on log–log paper. Since Eq. (8-12) is complex, the magnitude and phase are plotted separately. To do this we first look at the logarithm of the magnitude of Eq. (8-12):

$$\begin{aligned} \log|A_v(j\omega)| &= \log\left|\frac{1}{1+j\omega\tau}\right| \\ &= \log\frac{1}{|1+j\omega\tau|} \\ &= \log 1 - \log|1+j\omega\tau| \\ &= -\log|1+j\omega\tau| \end{aligned} \qquad (8\text{-}15)$$

To plot Eq. (8-15), we again study the low-frequency and the high-frequency values of this equation. At low frequencies for $\omega\tau < 1$, or $\omega < 1/\tau$, Eq. (8-15) becomes

$$\log|A_v(j\omega)| = 0 \qquad (8\text{-}16)$$

At high frequencies for $\omega\tau > 1$, or $\omega > 1/\tau$, Eq. (8-15) becomes

$$\begin{aligned} \log|A_v(j\omega)| &= -\log\omega\tau \\ &= -(\log\omega + \log\tau) \end{aligned} \qquad (8\text{-}17)$$

The value of $\log|A_v|$ can be plotted quickly as a function of $\log\omega$, since both Eqs. (8-16) and (8-17) are equations of straight lines. First, Eq. (8-16) is plotted for all values of ω up to $1/\tau$, since this is the range of frequencies for which it is valid. Equation (8-16) is of course a straight line along the horizontal axis. Equation (8-17) is plotted for all values of ω greater than $1/\tau$. Notice that this equation is a straight line of a unity slope that intersects the horizontals axis at $\omega = 1/\tau$, The resulting lines are shown in Fig. 8-7 along with the actual curve. The two straight lines form approximations

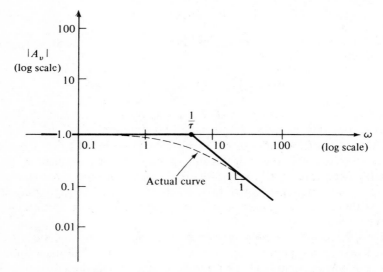

FIGURE 8-7 Plot of $|A_v|$ versus frequency on log-log paper.

to the actual curve with the largest deviation occurring at frequency $1/\tau$. This frequency is known as the *corner* or *break frequency*. Notice that there is fairly good agreement at all the frequencies, except in the neighborhood of the break frequency.

The phase of the gain function in Eq. (8-13) is simply

$$\phi = -\tan^{-1} \omega\tau \tag{8-18}$$

This is generally plotted as a function of log ω and is simply the arctangent function. The resulting curve is shown in Fig. 8-8. At the low frequencies, the value of the phase is zero; at the high frequencies it is $90°$; and at the break frequency it is $45°$. Generally speaking, in electronics, a plot of the phase function is not often done, since in usual applications the gain is of greater importance.

8.5 Bode Frequency-Response Plots

Having seen the method for obtaining the frequency-response plots for a simple *RC* network, let us now extend this method to a larger class of networks, such as those found in electronic circuits. The general network func-

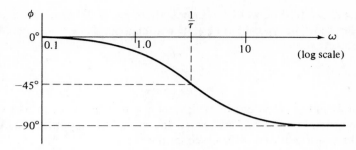

FIGURE 8-8 A plot of ϕ versus frequency on semi-log paper.

tion can usually be written as a ratio of two polynomials in s having the general form

$$G(s) = \frac{a_0 s^n + a_1 s^{n-1} + \ldots + a_n}{b_0 s^m + b_1 s^{m-1} + \ldots + b_m} \tag{8-19}$$

If the two polynomials are factored, we obtain

$$G(s) = \frac{a_0(s - z_1)(s - z_2)\ldots(s - z_n)}{b_0(s - p_1)(s - z_2)\ldots(s - p_m)} \tag{8-20}$$

where $z_1, z_2, \ldots, z_n$ are the *zeros* of network function, and $p_1, p_2, \ldots, p_m$ are the *poles* of the network function. In general, the poles and zeros may be real, purely imaginary, or complex.

Let us only assume that the poles and zeros are real and nonrepeated. Equation (8-20) can then be written

$$G(s) = \frac{a_0 \left(s + \dfrac{1}{T_1}\right)\left(s + \dfrac{1}{T_3}\right)\cdots}{b_0 \left(s + \dfrac{1}{T_2}\right)\left(s + \dfrac{1}{T_4}\right)\cdots} \tag{8-21}$$

where

$$T_1 = -\frac{1}{z_1}, \quad T_3 = -\frac{1}{z_2}, \quad \ldots$$

and

$$T_2 = -\frac{1}{p_1}, \quad T_4 = -\frac{1}{p_2}, \quad \ldots$$

Equation (8-21) can now be written more conveniently in the form

$$G(s) = \frac{a_0 T_2 T_4 \cdots}{b_0 T_1 T_3 \cdots} \frac{(1 + sT_1)(1 + sT_3)\cdots}{(1 + sT_2)(1 + sT_4)\cdots} \tag{8-22}$$

The logarithm of the magnitude of $G(j\omega)$ becomes

$$\begin{aligned}\log|G(j\omega)| = {}&\log \frac{a_0 T_2 T_4 T_6 \cdots}{b_0 T_1 T_3 T_5 \cdots} \\ &+ \log|1 + j\omega T_1| + \log|1 + j\omega T_3| + \ldots \\ &- (\log|1 + j\omega T_2| + \log|1 + j\omega T_4| + \ldots)\end{aligned} \tag{8-23}$$

By doing this, all the terms in the summation of Eq. (8-23) have the same form except the first. The typical term can be written as

$$\log|1 + j\omega T_i| = \log|G_i|$$

where

$$G_i = 1 + j\omega T_i \tag{8-24}$$

A study of the graph of this typical term would enable us to plot a complete frequency-response curve, since the complete response would be a simple graphical addition of such equations, that is,

$$\log|G(j\omega)| = \log \frac{a_0 T_2 T_4 T_6}{b_0 T_1 T_3 T_5} \\ + \log|G_1| + \log|G_3| + \ldots \\ - (\log|G_2| + \log|G_4| + \ldots) \tag{8-25}$$

Extracting the typical term from Eq. (8-25), we obtain

$$\log|G_i| = \log|1 + j\omega T_i| \\ = \log(1 + \omega^2 T_i^2)^{1/2} \\ = \tfrac{1}{2}\log(1 + \omega^2 T_i^2) \tag{8-26}$$

For low frequencies, or $\omega T_i \ll 1$, Eq. (8-26) becomes

$$\log|G_i| = \tfrac{1}{2}\log(1) = 0 \tag{8-27}$$

and for high frequencies, or $\omega T_i \gg 1$, Eq. (8-26) becomes

$$\log|G_i| = \tfrac{1}{2}\omega^2 T_i^2 = \log \omega T_i \\ = \log \omega + \log T_i \tag{8-28}$$

We recognize Eq. (8-28) to a be straight line with a positive slope of 1 and having an intercept of $\log T_i$ on the $\log|G_i|$ axis.

Equations (8-27) and (8-28) constitute the asymptotic behavior of $|G_i|$ and can be plotted on log–log paper, as shown in Fig. 8-9. The frequency $\omega = 1/T_i$ is called the *break frequency*, the *corner frequency*, or *break point*. We must recognize that the graph of Fig. 8-9 is only an approximation of the actual curve. For example, at the frequency $1/T_i$ we can calculate the value of the actual function from Eq. (8-26):

$$|G_i|_{\omega=1/T_i} = |(1 + j\omega T_i)|_{\omega=1/T_i} \\ = \left[1^2 + \left(\frac{1}{T_i}T_i\right)^2\right]^{1/2} \tag{8-29} \\ = \sqrt{2} \simeq 1.414$$

If $|G_i|$ is calculated as logarithmic gain in dB, we would obtain

$$20\log|G_i| = 20\log\sqrt{2} \\ \simeq 3 \text{ dB} \tag{8-30}$$

This means that the actual curve, as sketched in Fig. 8-9, is 3 dB larger than,

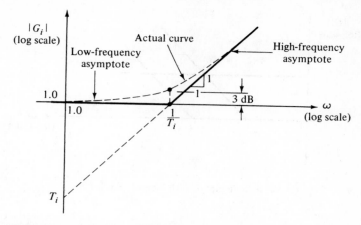

FIGURE 8-9 The Bode magnitude plot of a typical term $|1 + j\omega T_i|$.

or 1.414 times as great as, the asymptotic approximation at the break frequency. By substitution of various other values for ω in Eq. (8-29), it can be seen that the largest error occurs at the break frequency.

To make a complete plot of the magnitude of $|G(j\omega)|$ for Eq. (8-19) as a function of ω, we see that we need only to sum graphically curves of the type shown in Fig. 8-9, as well as the constant term.

Example 8-1: A network function is given by

$$G(j\omega) = \frac{3\left(1 + j\frac{\omega}{2}\right)}{1 + j\frac{\omega}{4}}$$

Sketch a Bode plot of the magnitude of $G(j\omega)$ as a function of ω.

Solution:

We recognize a constant term, 3, in the equation that plots as a horizontal straight line, as a function of frequency. We also recognize two break frequencies, 2 and 4. The factors of $G(j\omega)$ are plotted in Fig. 8-10a and are summed appropriately in Fig. 8-10b. Notice that the denominator term is actually subtracted graphically, resulting in a one-to-one slope between $\omega = 2$ and $\omega = 4$, and then in a zero slope, since the effect of the numerator and denominator terms cancel for $\omega > 4$.

Example 8-2: A network function is given by

$$G(j\omega) = \frac{2j\omega}{1 + j\frac{\omega}{3}}$$

Sketch a Bode plot of $|G(j\omega)|$ versus ω.

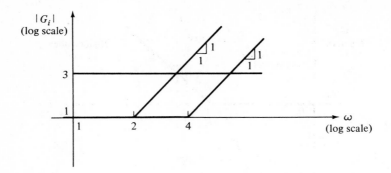

(a) A Bode plot of each factor in Ex. 8-1

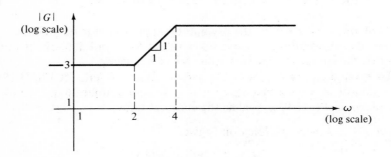

(b) The complete Bode plot for the network function in Ex. 8-1

FIGURE 8-10 Solution curves for Ex. 8-1.

Solution:

We recognize this problem as being somewhat different from the first, since the numerator does not contain the typical term studied in this section. However, it is easy to see that the factor $j\omega$ is merely a straight line through the origin that is added to the other two terms: the constant and the denominator factor with a break frequency $\omega = 3$ rad/s. The resulting Bode plot is shown in Fig. 8-11.

Very frequently, network functions such as voltage gain are measured in dB. This is very convenient, since many ac laboratory meters are calibrated this way. Hence plots of $|G|$ in dB versus log ω are made in exactly the same way, except that the one-to-one slope becomes a 20-dB-per-decade or 6-dB-per-octave slope. The advantages are that the experimenter needs only semilogarithmic paper to record his data, and the corner frequencies can be easily found by noting the frequencies at which the gain changes by 3 dB.

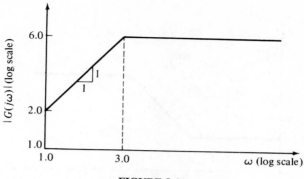

FIGURE 8-11

PROBLEMS

8.1 If I_{CEO} for a germanium transistor is measured to be 60 μA and β is found to be 30, what is the value of I_{CBO}?

8.2 (a) If the voltage gain of amplifier is 10, what is it in dB?
 (b) If it is 100, what will it be in dB?
 (c) 1,000?
 (d) 200?

8.3 Sketch the magnitude Bode plots of the network functions
 (a) $\dfrac{1}{s+a}$
 (b) $s+b$
 (c) $\dfrac{s+a}{s+b}$, where $a > b$
 (d) $\dfrac{s+a}{s+b}$, where $b > a$
 (e) $\dfrac{1+4s}{1+6s}$
 (f) $\dfrac{600(1+0.001s)}{(1+0.005s)}$

8.4 (a) Obtain the value of χ in Fig. P8-4.
 (b) Obtain a general formula for finding χ for Bode plots of the type shown in Fig. P8-4, when ω_1, ω_2, and y are known.

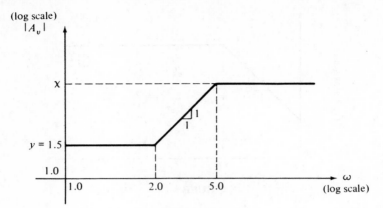

FIGURE P8-4

8.5 Sketch the magnitude Bode plots of the network functions
(a) s
(b) $\dfrac{1}{s}$
(c) $\dfrac{s(1 + 6s)}{1 + 2s}$
(d) $\dfrac{s(1 + 0.001s)}{1 + 0.005s}$
(e) $\dfrac{s(s + 30)}{(s + 40)(s + 100)}$

8.6 Obtain the magnitude Bode plots for the voltage gain v_o/v_i of the networks in Fig. P8-6.
(a)
(b)
(c)

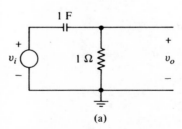

(a)

FIGURE P8-6

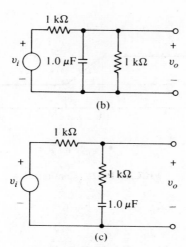

FIGURE P8-6 (Continued)

8.7 Find the value of x and y in each of the Bode plots shown in Fig. P8-7 by using simple geometry.
(a)
(b)
(c)

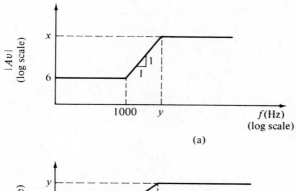

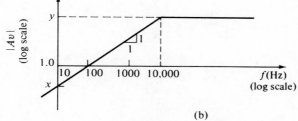

FIGURE P8-7

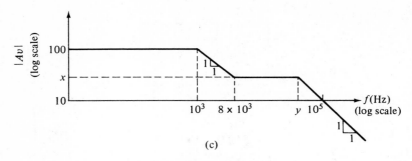

FIGURE P8-7 (Continued)

REFERENCES

[1] CUTLER, P., *Semiconductor Circuit Analysis* (McGraw-Hill Book Co., New York, 1964.)

[2] CHENG, D. K., *Analysis of Linear Systems* (Addison-Wesley Publishing Co., Inc., Reading, Mass., 1959).

[3] SESHU, S., and BALBANIAN, N., *Linear Network Analysis* (John Wiley & Sons, New York, 1959).

[4] VAN VALKENBURG, M. E., *Network Analysis*, 2nd ed. (Prentice-Hall, Inc., Englewood Cliffs, N.J., 1964).

9

LOW-FREQUENCY RESPONSE OF TRANSISTOR AMPLIFIERS

The previous chapters assumed that all coupling and bypass capacitors were large. That is, they were assumed large enough to be considered a short circuit to voltage variations when the small-signal equivalent circuit was drawn. The question that remains to be answered is: How large must these capacitors be in order that they are indeed short circuits? The actual value of the capacitor, of course, depends upon relative values of the impedance in the circuit as well as the frequency at which the circuits will operate. A more complete answer can be found in a study of the frequency response of the networks, which is obtained easily by the use of Bode plots, discussed in Chapter 8.

9.1 High-Pass R–C Filter

Consider the network shown in Fig. 9-1. This type of network is usually called a *high-pass filter*, since it is capable of passing high frequencies when C is practically a short circuit, and attenuating the low frequencies when the reactance of C becomes large. The circuit has been broken into three parts—with v_i and R_s representing the signal source, C and R representing the filter, and v_o representing the output signal.

By use of the voltage division method, the gain can be expressed as

$$\frac{V_o}{V_i} = \frac{R}{R + R_s + \frac{1}{sC}} = \frac{sRC}{1 + sC(R + R_s)} \qquad (9\text{-}1)$$

The asymptotic frequency-response curve can then be sketched, as shown

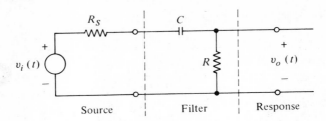

FIGURE 9-1 An R–C high-pass filter.

in Fig. 9-2. It is apparent from Fig. 9-2 why the network is called a high-pass filter. At frequencies greater than ω_1, the critical frequency, the gain remains almost constant at $R/(R + R_s)$. At frequencies below ω_1 the gain decreases as shown by the one-to-one slope, or at the rate of 6 dB per octave.

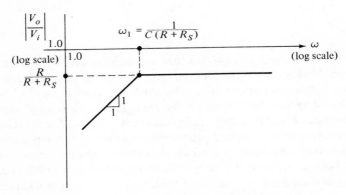

FIGURE 9-2 Asymptotic frequency response of the high-pass filter.

The observations made on the frequency response of the network of Fig. 9-1 may also be considered in a different manner. The following question can now be asked: What value of C is needed for the gain of the network to be $R/(R + R_s)$, if the lowest frequency to be used in $v_i(t)$ is ω_1?

If a 3-dB error in gain is not considered to be too large, but larger errors than 3 dB are unacceptable, then the answer is that the value of C can be given as

$$C \geq \frac{1}{\omega_1(R + R_s)} \quad (9\text{-}2)$$

The greater-than sign is used in Eq. (9-2), since a choice of capacitor larger than $1/\omega_1(R + R_s)$ is acceptable because the capacitor will pass frequencies above ω_1 with even less error than 3 dB in the voltage gain. The frequency ω_1 is usually known as the *cutoff frequency* of the high-pass filter.

9.2 Input Coupling Capacitor

The same analysis procedure used in Section 9.1 may now be applied to determine the size of a coupling capacitor in a transistor amplifier. Consider the amplifier shown in Fig. 9-3a. The approximate equivalent circuit is obtained in the usual way. The capacitor C_i is included, however, since its effect is to be analyzed.

The input current I_b becomes

$$I_b = \frac{V_i}{h_{ie} + \dfrac{1}{sC_i} + R_S} \tag{9-3}$$

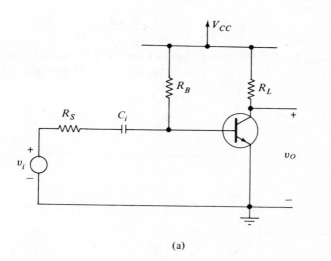

(a)

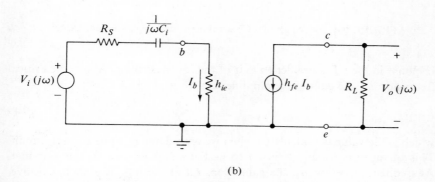

(b)

(a) CE amplifier with a single coupling capacitor C_i
(b) The transformed equivalent circuit of the CE amplifier

FIGURE 9-3 The CE amplifier.

and the output voltage is given by

$$V_o = -h_{fe}I_bR_L \tag{9-4}$$

Substitution of Eq. (9-3) into Eq. (9-4) yields

$$V_o = \frac{-h_{fe}R_LV_i}{h_{ie} + \dfrac{1}{sC_i} + R_S} \tag{9-5}$$

The voltage gain thus becomes

$$A_v = \frac{V_o}{V_i} = \frac{-h_{fe}R_L}{h_{ie} + \dfrac{1}{sC_i} + R_S} = \frac{-h_{fe}R_L sC_i}{1 + sC_i(R_S + h_{ie})} \tag{9-6}$$

and

$$A_v(j\omega) = \frac{-h_{fe}R_L j\omega C_i}{1 + j\omega C_i(R_S + h_{ie})} \tag{9-7}$$

The magnitude of the voltage gain of Eq. (9-7) can now be plotted on a Bode diagram, as shown in Fig. 9-4. Notice the similarity of the frequency

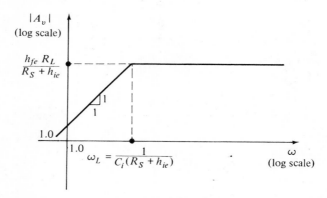

FIGURE 9-4 Bode plot of a CE amplifier with a single coupling capacitor.

response in the CE amplifier to that of the low-pass filter. Above the critical frequency, ω_L the gain of the amplifier is

$$A_v\big|_{\omega > \omega_L} = \frac{-h_{fe}R_L}{h_{ie} + R_S} \tag{9-8}$$

which is the value we calculated when we considered C_i to be a short circuit. This quantity is usually referred to as the *mid-band gain* of the amplifier. At frequencies below ω_L, the gain drops off at the rate of 6 dB per octave. At ω_L the gain is 3 dB below the mid-band gain value.

The frequency ω_L is usually known as the *low-frequency cutoff* of the amplifier. In general, the low-frequency cutoff of an amplifier is defined as that frequency that is 3 dB below the mid-band value when the frequency is decreased below the mid-band frequencies.

There is also a *high-frequency cutoff*, which is the frequency at which the gain decreases to 3 dB below the mid-band gain level as the frequency is increased. The phenomenon will be studied in a later chapter.

In practical design problems a maximum value of the lower cutoff frequency is usually specified. For example, in audio amplifiers the lower cutoff frequency is usually not any larger than 20 Hz. From this information we can set a minimum value for C_i, that is,

$$C_i \geq \frac{1}{\omega_L(R_S + h_{ie})} \qquad (9\text{-}9)$$

where $\qquad \omega_L =$ the required lower cutoff frequency

The greater-than sign in Eq. (9-9) is used, since a choice of C_i larger than $1/\omega_L(R_S + h_{ie})$ would make the deviation of the gain less than 3 dB for the required frequency range.

Example 9-1: Calculate the minimum value of C_i for the amplifier shown in Fig. 9-3a. The lower cutoff frequency f_L is, at most, 20 Hz. R_S is 1.0 kΩ, and h_{ie} is 2 kΩ.

Solution:

$$\omega_L = 2\pi f_L = 2\pi \times 20 = 125.6 \text{ rad/s}$$

$$C_i \geq \frac{1}{\omega_L(R_S + h_{ie})} = \frac{1}{1.25 \times 10^2 \times 3.0 \times 10^3}$$
$$= 2.7 \times 10^{-6} \text{ F}$$
$$= 2.7 \ \mu\text{F}$$

The minimum value of C_i is 2.7 μF.

9.3 Emitter Bypass Capacitor

In the *H*-type CE amplifier the size of the emitter bypass capacitor is very important, since its function is to short the emitter to ground in an ac sense. Figure 9-5a illustrates this type of amplifier, and Fig. 9-5b shows the approximate equivalent circuit. The coupling capacitor C_i is chosen to be very large so that our analysis can concentrate on the effect of C_E, the bypass capacitor. In Fig. 9-5c is a Thévenin equivalent of the input source, the source resistance, and the biasing resistors (the part of the circuit to the left of the dashed lines in Fig. 9-5b). In Fig. 9-5c, the following symbols are used:

$$R_G = R_S \| R_A \| R_B \qquad (9\text{-}10)$$

$$v_g = \frac{R_A \| R_B}{R_S + (R_A \| R_B)} v_i \qquad (9\text{-}11)$$

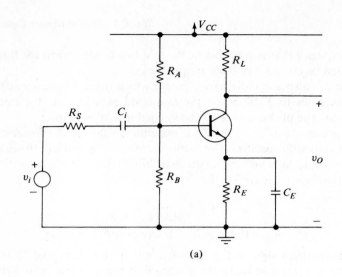

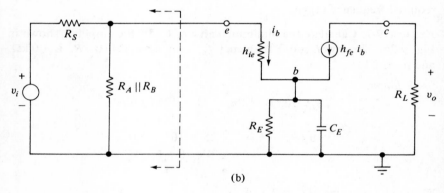

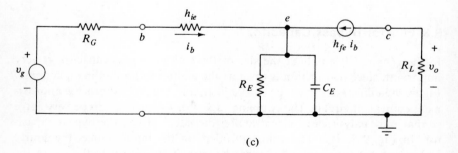

(a) A *CE* amplifier with a bypass capacitor and a very large coupling capacitor C_i.
(b) Approximate equivalent circuit
(c) Simplified equivalent circuit

FIGURE 9-5 CE amplifier and equivalent circuit.

Since we are concerned with the shape of the frequency-response curve and the break frequencies, we shall calculate the voltage gain A'_v, where

$$A'_v = \frac{v_o}{v_g} \tag{9-12}$$

remembering that, since v_g is proportional to v_i, the actual voltage gain $A_v = v_o/v_i$ will differ only by a constant and will be proportional to A'_v.

In Fig. 9-5c, the transform of the output voltage is given by

$$V_o = -\beta I_b R_L \tag{9-13}$$

and the transform of the input voltage becomes

$$V_g = (R_G + h_{ie})I_b + (\beta + 1)I_b \frac{R_E}{1 + sC_E R_E} \tag{9-14}$$

The voltage gain A'_v becomes

$$\begin{aligned} A'_v &= \frac{-\beta R_L}{(R_G + h_{ie}) + \dfrac{(\beta + 1)R_E}{1 + sC_E R_E}} \\ &= \frac{-\beta R_L(1 + sC_E R_E)}{[(\beta + 1)R_E + R_G + h_{ie}]\left[1 + s\dfrac{C_E R_E(R_G + h_{ie})}{(\beta + 1)R_E + R_G + h_{ie}}\right]} \\ &= \frac{-\beta R_L\left(1 + \dfrac{s}{\omega_2}\right)}{[(\beta + 1)R_E + R_G + h_{ie}]\left(1 + \dfrac{s}{\omega_1}\right)} \end{aligned} \tag{9-15}$$

where

$$\omega_2 = \frac{1}{C_E R_E}$$

and

$$\omega_1 = \frac{(\beta + 1)R_E + R_G + h_{ie}}{C_E R_E(R_G + h_{ie})}$$

The expression for ω_1 can be written in a more convenient form as

$$\omega_1 = \frac{\beta + 1}{C_E} \frac{(\beta + 1)R_E + (R_G + h_{ie})}{(\beta + 1)R_E(R_E + h_{ie})} \tag{9-16}$$

This expression may be simplified, since, in many cases, $(\beta + 1)R_E$ is much larger than $R_G + h_{ie}$

$$\begin{aligned} \omega_1 &\simeq \frac{(\beta + 1)R_E}{C_E R_E(R_G + h_{ie})} \\ &\simeq \frac{1}{C_E\left(\dfrac{R_G}{\beta} + r_e\right)} \end{aligned} \tag{9-17}$$

The magnitude of the gain A'_v in Eq. (9-15) may be plotted on a Bode diagram as shown in Fig. 9-6. It is apparent from Eqs. (9-15) and (9-17) that ω_2 is smaller than ω_1.

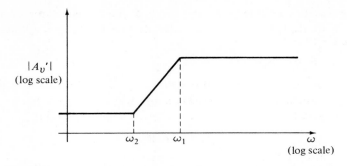

FIGURE 9-6 Bode plot indicating the shape of the frequency response due to the bypass capacitor in a CE amplifier.

The frequency ω_1 becomes approximately the lower cutoff frequency of the amplifier. This is not exactly a "3-dB-down point," since there is a contribution to the gain from the numerator term. Thus, to design for the bypass capacitor, we specify the inequality

$$C_E \geq \frac{1}{\omega_L\left(r_e + \dfrac{R_G}{\beta}\right)} \tag{9-18}$$

where $\quad R_G = R_A \| R_B \| R_S$

and $\quad \omega_L = $ the required lower cutoff frequency

In the case where R_G is zero, or the amplifier is driven by an ideal voltage source, it is easy to see that the value of C_E may be given as

$$C_E \geq \frac{1}{\omega_L r_e} \tag{9-19}$$

where $\omega_L = $ the required lower cutoff frequency

Example 9-2:
(a) Calculate the value of C_E for the amplifier shown in Fig. 9-7, if the lower cutoff frequency is to be no more than 20 Hz. Assume C_i to be infinitely large.
(b) Sketch the Bode frequency-response plot for the gain.

Solution:

(a) The Thévenin equivalent impedance for the input circuit R_G becomes

$$R_G = R_S \| R_A \| R_B = 720 \, \Omega$$

$$r_e = \frac{m v_T}{I_E} = \frac{(1.4)(0.026)}{1 \times 10^{-3}} = 36.4 \, \Omega$$

$$C_E > \frac{1}{\omega_L\left(r_e + \dfrac{R_G}{\beta}\right)} = \frac{1}{40\pi(43.6)} = 182 \, \mu\text{F}$$

(We use a 200-μF electrolytic capacitor.)

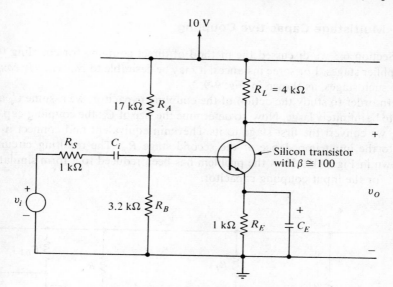

FIGURE 9-7 Circuit for Ex. 9-2.

(b) The two break frequencies may now be calculated:

$$\omega_1 \simeq \frac{1}{C_E\left(r_e + \frac{R_G}{\beta}\right)}$$

$$= \frac{1}{200 \times 43.6 \times 10^{-6}} = 115 \text{ rad/s} \simeq 18 \text{ Hz}$$

$$\omega_2 = \frac{1}{R_E C_E}$$

$$= \frac{1}{10^3 \times 2 \times 10^{-4}} = 5 \text{ rad/s} \simeq 1 \text{ Hz}$$

Figure 9-8 shows the resulting Bode plot.

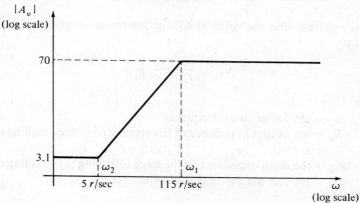

FIGURE 9-8 Solution to Ex. 9-2.

9.4 Multistage Capacitive Coupling

In Section 6.3 we discussed the method of direct coupling for coupling two amplifier stages. For some instances, it may be desirable to *capacitively couple* two such stages, as shown in Fig. 9-9.

In order to study the action of the coupling capacitor, we assume C_i and C_E to be infinitely large. Now, to determine the size of C_c, the coupling capacitor, we convert the first stage to its Thévenin equivalent and connect it via C_c to the input impedance of the second stage R_{in}. The resulting circuit is shown in Fig. 9-10. Now the problem has been reduced to a form similar to that for the input coupling capacitor.

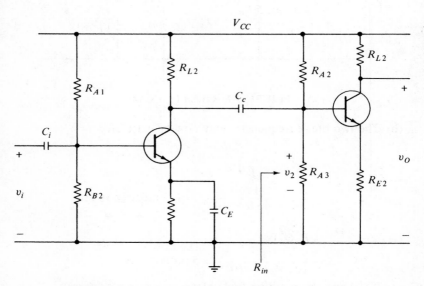

FIGURE 9-9 A two-stage amplifier capacitively coupled.

It is apparent that the value of C_c, the interstage coupling capacitor, is given by

$$C_c \geq \frac{1}{\omega_L(R_o + R_{in})} \tag{9-20}$$

where,

ω_L = the lower cutoff frequency
R_o = the output impedance of the stage driving the coupling capacitor
R_{in} = the input impedance of the stage following the coupling capacitor (or, the external load of the stage)

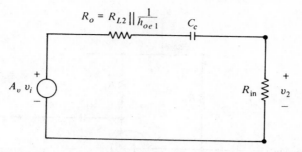

FIGURE 9-10 An equivalent circuit for the two-stage amplifier of Fig. 9-9.

There are many other cases in which the size of the capacitor is critical upon the low-frequency response of an amplifier stage. It is left as an exercise for the reader to work out the design equations as the need for them arises. The examples in this and the previous sections should help to show the method and the type of reasoning involved. The following are a few of these cases:

1. The output coupling capacitor in an emitter follower.
2. The input and output coupling capacitors of a common base amplifier.
3. The bypass capacitor in a common base amplifier.
4. The capacitors in a bootstrapped amplifier stage.

9.5 Combined Effects of the Two Coupling Capacitors

Let us now consider the effect of two capacitors simultaneously, without assuming one as infinitely large. The example of the current biased amplifier with its two coupling capacitors is chosen purposely to simplify the analysis. The amplifier is illustrated in Fig. 9-11a and the approximate equivalent circuit in Fig. 9-11b. Both coupling capacitors are shown, since we are considering them simultaneously.

From Fig. 9-11b the transform of the output voltage is

$$V_o = -h_{fe}I_b \frac{R_L\left(\frac{1}{sC_o} + R_{ext}\right)}{R_L + \frac{1}{sC_o} + R_{ext}}$$

$$= \frac{-h_{fe}I_b R_L(1 + sC_o R_{ext})}{1 + sC_o(R_L + R_{ext})} \quad (9\text{-}21)$$

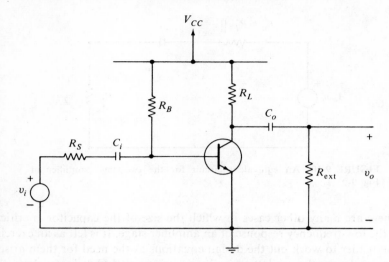

(a) A simple CE amplifier with two coupling capacitors.

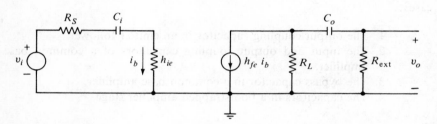

(b) The equivalent circuit with the two capacitors.

FIGURE 9-11 CE amplifier with two coupling capacitors and equivalent circuit.

and the transform of the input voltage is

$$V_i = I_b\left(R_S + \frac{1}{sC_i} + h_{ie}\right)$$
$$= I_b \frac{1 + sC_i(R_S + h_{ie})}{sC_i} \tag{9-22}$$

Dividing Eq. (9-21) by Eq. (9-22) we obtain

$$A_v(s) = \frac{-h_{fe}R_L sC_i(1 + sC_o R_{ext})}{[1 + sC_i(R_S + h_{ie})][1 + sC_o(R_L + R_{ext})]} \tag{9-23}$$

Equation (9-23) can now be rewritten as follows:

$$A_v(s) = \frac{-h_{fe}R_L sC_i\left(1 + \dfrac{s}{\omega_2}\right)}{\left(1 + \dfrac{s}{\omega_1}\right)\left(1 + \dfrac{s}{\omega_3}\right)} \tag{9-24}$$

where
$$\omega_1 = \frac{1}{C_i(R_S + h_{ie})}$$

$$\omega_2 = \frac{1}{C_o R_{ext}}$$

and
$$\omega_3 = \frac{1}{C_o(R_L + R_{ext})}$$

Notice that ω_2 is always larger than ω_3.

If Eqs. (9-9) and (9-20) are now employed with the equality sign rather than the greater-than sign to determine C_i and C_o, then it follows that ω_1 is equal to ω_3, and the Bode plot for the voltage gain, shown in Fig. 9-12, results.

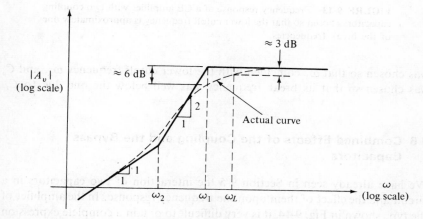

FIGURE 9-12 A possible frequency response for the CE amplifier with two coupling capacitors only.

If ω_2 is much less than ω_1 and ω_3 in the figure, then the gain at ω_1 drops off approximately 6 dB, and therefore the lower cutoff frequency ω_L would be somewhat higher than ω_1.

Thus it is very easy to see why the greater-than sign should be imposed on Eqs. (9-9) and (9-20) for design purposes if more than one coupling capacitor is used in the circuit.

In trying to solve the problem of the interaction of the two capacitors on the gain at low frequencies, the designer has one of two choices. First, he can make both capacitors 3 or 4 times larger than they need to be to ensure that the lower cutoff frequency will be much lower than required. Second, he may choose to use Eqs. (9-9) and (9-20) with the equal sign and make one of the capacitors 10 times larger than it should be, forcing one of the frequencies far enough away so that it will have no effect on the required lower cutoff frequency. This latter choice is shown in Fig. 9-13. Here the capacitor C_i

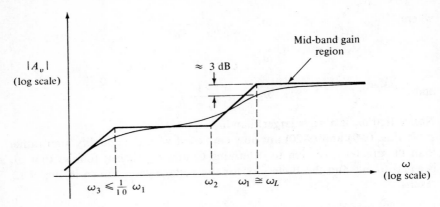

FIGURE 9-13 Frequency response of a CE amplifier with two coupling capacitors chosen so that the lower cutoff frequency is approximately one of the break frequencies.

was chosen so that ω_1 coincided with the lower cutoff frequency ω_L, and C_o was chosen so that its break frequency was well below the one due to C_i.

9.6 Combined Effects of the Coupling and the Bypass Capacitors

We have already seen in Section 9-5 the interaction of two capacitors in a circuit and the effect of them upon the frequency response. In the amplifier of the type shown in Fig. 9-14, it is very difficult to obtain a complete expression for the voltage gain with all three capacitors in the equivalent circuit. Even if this is achieved, it is also very difficult to establish design equations to give a 3 dB reduction in the gain for a specified lower frequency cutoff. Thus we employ an *approximate* solution to determine the size of the capacitors.

Let us list the design equations previously obtained for the capacitors and apply them to the circuit of Fig. 9-14:

$$C_i \geq \frac{1}{\omega_L(R_S + R_{in})} \tag{9-25}$$

$$C_o \geq \frac{1}{\omega_L\left(R_L \middle\| \frac{1}{h_{oe}} + R_{ext}\right)} \tag{9-26}$$

$$C_E \geq \frac{1}{\omega_L\left(r_e + \frac{R_G}{\beta}\right)} \tag{9-27}$$

where
$$R_{in} = h_{ie} \| R_A \| R_B$$

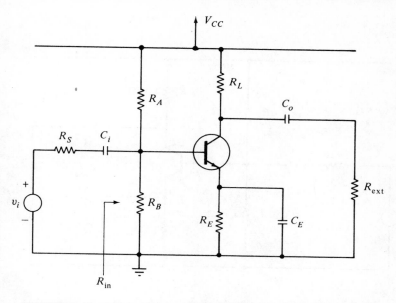

FIGURE 9-14 A CE amplifier with three capacitors to be considered for design.

If the equality sign is used on any of these three equations to obtain ω_L and the other capacitors are made *much* larger than this value, say 10 times, then the 3-dB-down break frequency occurs at ω_L, since the remaining two could be considered to be short circuits at this frequency.

Upon examination of these three design equations, it is apparent that C_E is probably the largest capacitor. Since the cost of capacitors increases appreciably with size, it would be reasonable to design C_E to produce the highest break frequency and the others to produce break frequencies well below this value. This procedure is illustrated by Example 9-3.

Example 9-3: The single stage amplifier shown in Fig. 9-15 is connected with a coupling capacitor C_o to a voltmeter, which is shown as an external load R_{ext}. T_1 is a silicon transistor. The emitter current I_E is found to be 1 mA. Complete the design of the amplifier by determining values of C_i, C_o, and C_E so that the lower cutoff frequency is 20 Hz at most.

Solution:

We calculate the value of C_E to be

$$C_E \geq \frac{1}{\omega_L\left(r_e + \dfrac{R_G}{\beta}\right)}$$

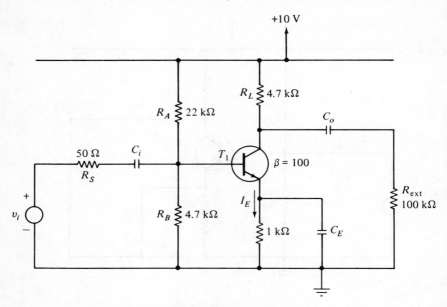

FIGURE 9-15 Circuit of Ex. 9-3.

Since $I_E = 1$ mA and since T_1 is a silicon transistor, we shall assume that $m = 1.4$ so that

$$r_e = \frac{mv_T}{I_E} \simeq 36 \, \Omega$$

The equivalent source impedance, including the resistors, becomes

$$R_G = R_A || R_B || R_S$$
$$= 22 \text{ k}\Omega || 4.7 \text{ k}\Omega || 50 \, \Omega$$
$$\simeq 50 \, \Omega$$

and

$$\frac{R_G}{\beta} = \frac{50}{100} = 0.5 \, \Omega$$

Hence

$$C_E \geq \frac{1}{(2\pi)(20)[(36) + 0.5]} \simeq 198 \, \mu\text{F}$$

We should use a 200- or 100-μF capacitor.
Now, since

$$R_{in} = R_A || R_B || h_{ie} = 22 \text{ k}\Omega || 4.7 \text{ k}\Omega || 3.6 \text{ k}\Omega = 1.9 \text{ k}\Omega$$

then

$$C_i \geq \frac{1}{\omega_L(R_S + R_{in})} = \frac{1}{2\pi(20)(50 \, \Omega + 1.9 \text{ k}\Omega)} = 4.1 \, \mu\text{F}$$

$$C_o \geq \frac{1}{\omega_L(R_L + R_{ext})} = \frac{1}{2\pi(20)(4.7 + 100) \times 10^3} = 0.08 \, \mu\text{F}$$

Since the design value of C_E is much larger, we shall make C_i and C_o about 10 times larger than these calculated values. That is, let

$$C_i = 50 \ \mu\text{F}$$
$$C_o = 1.0 \ \mu\text{F}$$
and
$$C_E = 200 \ \mu\text{F}$$

It is interesting to note the relatively small value of C_o. This is primarily due to the relatively high resistance load connected to the amplifier.

9.7 Frequency Effects in an FET due to Coupling and Bypass Capacitors

The design of FET amplifiers for a specified low-frequency cutoff is similar to that of bipolar transistors. Consider the common source amplifier in Fig. 9-16a along with its equivalent circuit in Fig. 9-16b.

It is interesting to note at this point that if v_s is either purely sinusoidal, or is a signal containing no dc component, C_i is unnecessary and can be replaced by a short circuit, since the gate-to-ground voltage V_{GO} is zero. However, if this is not the case, it is easy to see that the value of C_i can be determined by

$$C_i \geq \frac{1}{\omega_L(R_i + R_g)} \qquad (9\text{-}28)$$

Equation (9-28) is easily obtained by assuming that C_s and C_o are short circuits, and by recognizing the fact that R_i, C_i, and R_g form a high-pass filter at the input.

In a similar manner, it is easy to obtain the design equation for C_o. The value of C_o is determined by

$$C_o \geq \frac{1}{\omega_L(R_L + R_{ext})} \qquad (9\text{-}29)$$

Equation (9-29) was also obtained by assuming the other two capacitors in the circuit to be short circuits.

The method for obtaining the bypass capacitor in an FET amplifier is similar to that of the bipolar transistor amplifier. The first step is to obtain the approximate equivalent circuit, assuming the coupling capacitors to be short circuits. This circuit is shown in Fig. 9-17. The next step is to calculate the voltage and plot the frequency response. To do this we shall designate the parallel equivalent of R_s and C_s, as Z_s,

$$Z_s = \frac{R_s}{1 + sC_sR_s} \qquad (9\text{-}30)$$

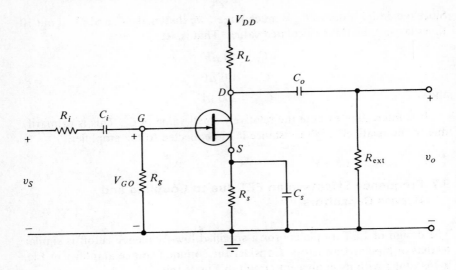

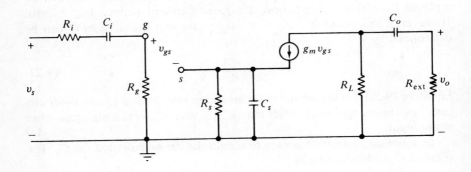

(a) A common source amplifier with three capacitors in the circuit.
(b) The small signal circuit of the common source amplifier for low frequency analysis.

FIGURE 9-16 Common source amplifier and low-frequency equivalent circuit.

and the parallel equivalent of R_L and R_{ext}, as R'_L, where

$$R'_L = \frac{R_L R_{ext}}{R_L + R_{ext}} \tag{9-31}$$

The transform of the output voltage is given by

$$V_o = -g_m V_{gs} R'_L \tag{9-32}$$

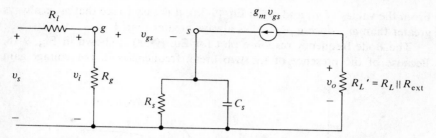

FIGURE 9-17 Small signal equivalent circuit of a common source amplifier necessary to determine a design equation for the source bypass capacitor.

and, using KVL at the input, we obtain

$$V_i = V_{gs} + g_m V_{gs} Z_s \tag{9-33}$$

where V_i is found, by voltage division, to be

$$V_i = \frac{R_g}{R_g + R_i} V_s \tag{9-34}$$

Using Eqs. (9-32), (9-33), and (9-34), the voltage gain becomes

$$\frac{V_o}{V_s} = \frac{-R_g}{R_g + R_i} g_m R_L \frac{1}{1 + g_m Z_s} \tag{9-35}$$

Equation (9-35) can be written more briefly as

$$\frac{V_o}{V_s} = \frac{-K}{1 + g_m Z_s} \tag{9-36}$$

where $\quad K =$ mid-band gain

Upon substituting the value of Z_s, obtained in Eq. (9-30), into Eq. (9-36), we obtain

$$\frac{V_o}{V_s} = \frac{-K}{1 + g_m \dfrac{R_s}{1 + sC_s R_s}} = \frac{-K(1 + sR_s C_s)}{(1 + g_m R_s)\left(1 + \dfrac{sR_s C_s}{g_m R_s + 1}\right)} \tag{9-37}$$

Equation (9-37) can be rewritten as

$$\frac{V_o}{V_s} = \frac{-K\left(1 + \dfrac{s}{\omega_1}\right)}{(1 + g_m R_s)\left(1 + \dfrac{s}{\omega_2}\right)} \tag{9-38}$$

where

$$\omega_1 = \frac{1}{R_s C_s}$$

and

$$\omega_2 = \frac{g_m R_s + 1}{R_s C_s}$$

From the values of ω_1 and ω_2 in Eq. (9-38), it is easy to see that ω_2 is always greater than ω_1, since $g_m R_s + 1$ is always greater than 1.

The Bode frequency response plot for Eq. (9-38) is shown in Fig. 9-18. Because of the presence of the two break frequencies in the voltage-gain

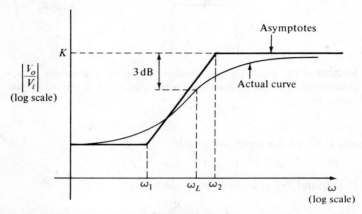

FIGURE 9-18 Frequency response of a common source amplifier due to the source bypass capacitor alone.

expression, it is apparent that ω_L, the lower cutoff frequency, is somewhat lower than ω_2. Rather than calculate this frequency and then obtain an equation for C_s, let us use ω_2 as the approximate lower cutoff frequency to find the value of C_s. This gives us a value of C_s that is slightly larger than necessary, since ω_2 is greater than ω_L. However, this is done anyway to ensure that specifications are met, and the design equation for determining the value of C_s becomes

$$C_s \geq \frac{1 + g_m R_s}{\omega_L R_s} \tag{9-39}$$

The design equations for the common source FET amplifier shown in Fig. 9-16a are listed in Table 9-1. Upon a close examination of this table and substitution of typical ranges of values for R_i, R_g, R_s, g_m, R_2, and R_{ext}, we see that C_s is usually the largest capacitor in the circuit, that C_i is much smaller than the input coupling capacitor of a bipolar transistor (since R_g, the input impedance, is so large—typically, 1 MΩ), and that C_o varies greatly depending upon the value of R_{ext}.

The design procedure here is also the same as in the bipolar transistor amplifier. Since C_s, the bypass capacitor, is the largest in the circuit, its value is determined by the minimum value given in Table 9-1. The coupling capacitors are made at least 10 times as large as the minimum value so as to move the other break frequencies to a much lower value.

Sec. 9.7 / Frequency Effects in an FET due to Coupling and Bypass Capacitors

TABLE 9-1 DESIGN EQUATIONS FOR CAPACITORS IN A CS–FET AMPLIFIER

Component	Minimum Value
C_i	$\dfrac{1}{\omega_L(R_i + R_g)}$
C_s	$\dfrac{1 + g_m R_s}{\omega_L R_s}$
C_o	$\dfrac{1}{\omega_L(R_L + R_{ext})}$

Example 9-4: Complete the design of the circuit shown in Fig. 9-19 by obtaining the values of the capacitors in the circuit. The FET has a g_m of 3 mA/V; the lower cutoff frequency is not to exceed 40 Hz.

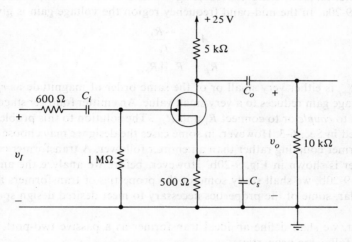

FIGURE 9-19 Circuit for Ex. 9-4.

Solution:

The minimum values of the capacitors are calculated by the formulas given in Table 9-1:

$$C_s \geq \frac{g_m R_s + 1}{\omega_L R_s} = \frac{(3 \times 10^{-3} \times 5 \times 10^2) + 1}{2\pi \times 40 \times 5 \times 10^2} = 20 \ \mu F$$

$$C_o \geq \frac{1}{\omega_L(R_L + R_{ext})} = \frac{1}{2\pi \times 40 \times 15 \times 10^3}$$
$$= 0.27 \times 10^{-6} \ F$$
$$= 0.27 \ \mu F$$

$$C_i \geq \frac{1}{\omega_L(R_i + R_g)} = \frac{1}{2\pi \times 40 \times (1 \times 10^6)}$$
$$= 0.004 \ \mu F$$

For our design values we shall choose C_s to be the minimum value and C_i and C_o to be at least 10 times larger than the minimum values so as to move their break frequencies well below 40 Hz. Typical values could be as follows:

$$C_s = 20 \ \mu\text{F (electrolytic capacitor)}$$
$$C_o = 5 \ \mu\text{F (electrolytic capacitor)}$$
and
$$C_i = 0.047 \ \mu\text{F}$$

9.8 Transformers for Transistor Circuits

In the previous sections we have seen how a transistor amplifier is coupled to an external load, using *RC coupling*. Consider the RC coupled amplifier in Fig. 9-20a. In the mid-band frequency region the voltage gain is given by

$$A_v = \frac{-R'_L}{r_e} \tag{9-40}$$

where
$$R'_L = R_L || R_{ext}$$

If R_{ext} is either very small or of the same order of magnitude as r_e, then the voltage gain reduces to a very small value. An emitter follower stage could be used to *couple* or to connect R_L and R_{ext}. The solution to this problem was discussed in Sec. 6-3. However, in some cases the designer may choose to use transformer coupling rather than an emitter follower. A transformer coupled amplifier is shown in Fig. 9-20b. However, before we analyze the amplifier of Fig. 9-20b, we shall study some of the properties of transformers and, in particular, some of the properties necessary to meet desired design specifications.

First, we shall define an ideal transformer as a passive two-port, which has the following properties:

$$v_2 = \frac{1}{a} v_1 \tag{9-41}$$

and
$$i_2 = a i_1 \tag{9-42}$$

where a is the turns ratio of the transformer.

The voltage and current reference directions for the ideal transformer are shown in Fig. 9-21a. The symbol used for the ideal transformer is shown in Fig. 9-21b. Notice that the word *ideal* is written underneath the transformer symbol to distinguish it from the symbol used for a pair of coupled coils.

There are two immediate properties of an ideal transformer that can be deduced from the defining Eqs. (9-41) and (9-42):

1. The product of v_2 and i_2 yields

$$v_2 i_2 = v_1 i_1 \tag{9-43}$$

or
$$p_1(t) = p_2(t) \tag{9-44}$$

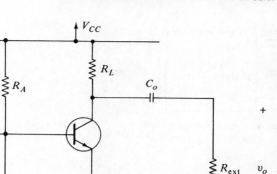

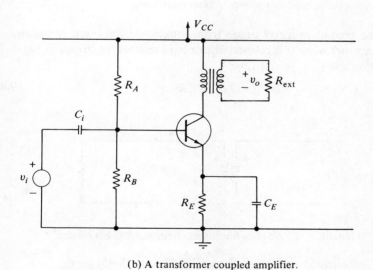

(a) An RC coupled amplifier.

(b) A transformer coupled amplifier.

FIGURE 9-20 A CE amplifier coupled to an external load.

where

$p_1 =$ the instantaneous input power delivered into the transformer

and $p_2 =$ the instantaneous output power delivered by the transformer

Since the instantaneous input power and the output power of an ideal transformer are equal, we say that the transformer is 100 per cent efficient.

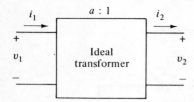

(a) A two-port representation of an ideal transformer.

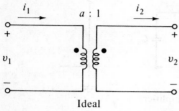

(b) The symbol for an ideal transformer.

FIGURE 9-21 Two representations for an ideal transformer.

2. The second property arises from looking at the input impedance of the transformer when it is connected to a load resistor, as shown in Fig. 9-22. Using Ohm's law, v_2 is given by

$$v_2 = i_2 R \qquad (9\text{-}45)$$

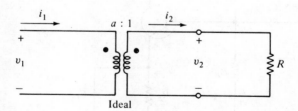

FIGURE 9-22 An ideal transformer terminated with a resistor R.

and the substitution of Eqs. (9-41) and (9-42) into (9-45) yields

$$\frac{1}{a} v_1 = a i_1 R \qquad (9\text{-}46)$$

Solving for the input impedance, we obtain

$$R_{in} = \frac{v_1}{i_1} = a^2 R \qquad (9\text{-}47)$$

This second property can be summarized as follows: The input impedance R_{in} of an ideal transformer that has the output connected to a load resistor

R is given by

$$R_{in} = a^2 R \qquad (9\text{-}48)$$

where a is the turns ratio of the ideal transformer.

Of course, as usual, there is nothing that is ideal in the real world. The two properties obtained for the ideal transformer, however, are very desirable, as shall be seen later. We shall now consider practical transformers in order that we may discover which properties a practical transformer must have to be approximately ideal. This is the same procedure that was considered in Chapter 1, where ideal and practical sources were compared.

A practical transformer usually consists of two coils of wire wound on a core, as shown in Fig. 9-23. The core material is often iron but may be any type of material. When no material is used at all, this type of transformer is known as an *air-core transformer*. The former type is known as an *iron-core transformer*.

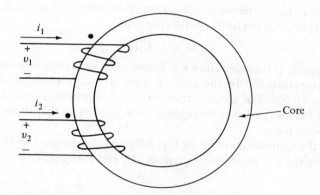

FIGURE 9-23 A practical or physical transformer showing the relationship between the dots and the winding direction.

The dots on a transformer are used to indicate the orientation of the coils in space. If the dots are placed in the manner shown in Fig. 9-23, then the equivalent circuit in Fig. 9-24 is used with the accompanying pair of equations to describe the practical transformer:

$$V_1 = (R_1 + sL_1)I_1 - sMI_2 \qquad (9\text{-}49)$$
$$-V_2 = -sMI_1 + (R_2 + sL_2)I_2 \qquad (9\text{-}50)$$

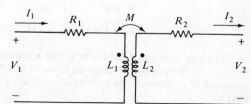

FIGURE 9-24 An equivalent circuit for a physical transformer.

where

R_1 = resistance of the input coil
R_2 = resistance of the output coil
L_1 = inductance of the input coil
L_2 = inductance of the output coil
M = the mutual inductance between the two coils

Equations (9-49) and (9-50) can be rewritten in matrix form as follows:

$$\begin{bmatrix} V_1 \\ -V_2 \end{bmatrix} = \begin{bmatrix} R_1 + sL_1 & -sM \\ -sM & R_2 + sL_2 \end{bmatrix} \begin{bmatrix} I_1 \\ I_2 \end{bmatrix}^* \quad (9\text{-}51)$$

In transformer theory a *coefficient of coupling* k is defined as

$$k = \frac{M}{\sqrt{L_1 L_2}} \quad (9\text{-}52)$$

Based on experimental observations it is found that k is always less than 1. In practical terms, k measures the *degree of coupling* or the *closeness* of the two coils. Hence we obtain the following result:

$$M \leq \sqrt{L_1 L_2} \quad (9\text{-}53)$$

The equality is included since k is found to be very close to unity in many practical transformers. In the case of some iron-core transformers, k can be as high as 0.998. For air-core transformers k approaches zero, as the coils are placed far away or at right angles to one another so that there is no coupling between them.

Using the equivalent circuit of Fig. 9-24 and the matrix Eq. (9-51), let us now impose the condition that V_1 equals aV_2 and I_1 equals I_2/a:

$$V_1 = aV_2 \quad (9\text{-}54)$$

$$I_1 = \frac{1}{a} I_2 \quad (9\text{-}55)$$

for all ω greater than ω_L. These equations are the properties of an ideal transformer. Note, however, that we shall impose these conditions only in the mid-band region, which are the frequencies greater than ω_L, the lower cutoff frequency. The only condition we shall impose on a is that it be a real constant.

To obtain Eqs. (9-54) and (9-55) from Eq. (9-51), we first assume R_1 and R_2 to be zero, or very small. When this is done, Eq. (9-51) reduces to

$$\begin{bmatrix} V_1 \\ -V_2 \end{bmatrix} = \begin{bmatrix} sL_1 & -sM \\ -sM & sL_2 \end{bmatrix} \begin{bmatrix} I_1 \\ I_2 \end{bmatrix} \quad (9\text{-}56)$$

*The sign convention used here is that when both the currents I_1 and I_2 either leave or enter at the dots, the off-diagonal terms in Eq. (9-51) are positive. Otherwise, they are negative. M is chosen to be positive always. See References [1] and [2] at the end of the chapter.

From Eq. (9-56) we obtain an expression for the voltage gain of the transformer as

$$\frac{V_2}{V_1} = \frac{-L_2 I_2 + M I_1}{L_1 I_1 - M I_2} \tag{9-57}$$

If an *iron-core transformer* is used, Eq. (9-57) becomes

$$\frac{V_2}{V_1} \simeq \frac{-L_2 I_2 + \sqrt{L_1 L_2}\, I_1}{L_1 I_1 - \sqrt{L_1 L_2}\, I_2} \tag{9-58}$$

since $M \simeq \sqrt{L_1 L_2}$ for $k \simeq 1$. Dividing the top on both sides by $\sqrt{L_2}$ and the bottom on both sides by $\sqrt{L_1}$, Eq. (9-58) becomes

$$\frac{V_2/\sqrt{L_2}}{V_1/\sqrt{L_1}} = \frac{-\sqrt{L_2}\, I_2 + \sqrt{L_1}\, I_1}{\sqrt{L_1}\, I_1 - \sqrt{L_2}\, I_2} = 1 \tag{9-59}$$

so that

$$\frac{V_2}{V_1} = \sqrt{\frac{L_2}{L_1}} \tag{9-60}$$

We note that Eq. (9-60) is of the same form as (9-54), if a is assumed to be $\sqrt{L_1/L_2}$.

Let us now list the assumptions that were made on the two coupled coils of Fig. 9-23 in order to obtain the voltage condition of an ideal transformer of Eq. (9-54):

1. The winding resistances R_1 and R_2 are zero, or negligible.
2. An iron-core transformer is used to obtain a coupling coefficient of unity.

There are only two assumptions!

To try to meet the condition imposed by Eq. (9-55), we shall now use a transformer with the above properties and connect it to a load resistor R as shown in Fig. 9-25. For the circuit in Fig. 9-25, the following matrix equation may now be written

$$\begin{bmatrix} V_1 \\ 0 \end{bmatrix} = \begin{bmatrix} sL_1 & -sM \\ -sM & sL_2 + R \end{bmatrix} \begin{bmatrix} I_1 \\ I_2 \end{bmatrix} \tag{9-61}$$

Using the second equation in (9-61), we obtain

$$s\sqrt{L_1 L_2}\, I_1 = (sL_2 + R) I_2 \quad \text{where } M = \sqrt{L_1 L_2} \text{ and } k = 1 \tag{9-62}$$

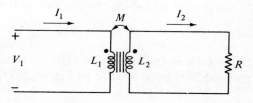

FIGURE 9-25 An idealized iron core transformer with a coupling coefficient of unity and zero winding resistance connected to a load resistance R.

which gives the following expression for current gain:

$$\frac{I_2}{I_1} = \frac{s\sqrt{L_1 L_2}}{sL_2 + R} = \frac{\sqrt{L_1 L_2}}{R} \frac{s}{1 + sL_2/R} \tag{9-63}$$

A Bode frequency-response plot, as shown in Fig. 9-26, illustrates the current gain as a function of frequency. Notice that in the mid-band frequency region, the current gain is given by

$$\frac{I_2}{I_1} = \sqrt{\frac{L_1}{L_2}} \qquad \text{for } \omega > \frac{R}{L_2} \tag{9-64}$$

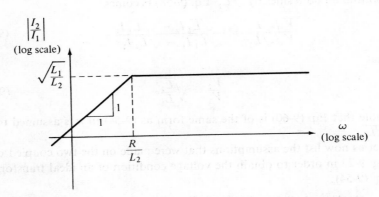

FIGURE 9-26 A Bode plot of the current gain of a practical transformer with negligible winding resistance.

A comparison of Eqs. (9-64) and (9-55) reveals the same form, if a is assumed to be $\sqrt{L_1/L_2}$. Notice that this is true for all frequencies greater than R/L_2. This result can be rewritten as follows: from Eq. (9-64):

$$\omega > \frac{R}{L_2} \tag{9-65}$$

or

$$\omega L_2 > R \tag{9-66}$$

Equation (9-66) imposes the final condition on the properties of a practical transformer. It can be stated as follows: A practical coupling transformer should have the reactance of the secondary greater than the load resistance connected to it for all frequencies for which the transformer will be used.

Since a, in a practical transformer, is given by

$$a = \sqrt{\frac{L_1}{L_2}} \tag{9-67}$$

let us look at the value of a, assuming that the transformer is wound in toroidal fashion, as implied by Fig. 9-23. The value of a toroidal inductance is given as

$$L = \frac{\mu A}{l} N^2 \qquad (9\text{-}68)$$

where

$\mu =$ permeability of the core material
$A =$ cross-sectional area of the core
$l =$ average length of the core material
$N =$ number of turns of wire the inductor is composed of

By substituting the values of L_1 and L_2, as given by Eq. (9-68), into Eq. (9-67), we see that a is indeed the turns ratio N_1/N_2:

$$a = \frac{N_1}{N_2} \qquad (9\text{-}69)$$

This is also approximately true for the more conventional *rectangular-shaped* core, since Eq. (9-68) is a good approximation for almost *any* shape of core, provided μ is large enough.

9.9 Analysis of Transformer Coupled Amplifiers

Let us consider the transformer coupled amplifier shown in Fig. 9-27a and its equivalent circuit shown in Fig. 9-27b. Assume the transformer to be ideal in the practical sense.

The voltage gain v_{ce}/v_i is found by the usual technique. Notice that, owing to the transformer, the effective load resistance between collector and emitter is R'_L, where

$$R'_L = a^2 R_L \qquad (9\text{-}70)$$

Thus

$$\frac{v_{ce}}{v_i} = \frac{-h_{fe} R'_L}{h_{ie}} = \frac{-h_{fe} a^2 R_L}{h_{ie}} \qquad (9\text{-}71)$$

Since the voltage gain across the transformer is given by

$$\frac{v_o}{v_{ce}} = -\frac{1}{a} \qquad (9\text{-}72)$$

the overall voltage gain is given as the product of Eqs. (9-71) and (9-72):

$$A_v = \frac{v_o}{v_i} = \frac{a h_{fe} R_L}{h_{ie}} \simeq \frac{a R_L}{r_e} \qquad (9\text{-}73)$$

It is apparent from Eq. (9-73) that if it is desirable to couple a small load resistance R_L to a transistor, then a should be a number larger than 1 so that the overall gain is increased. Consequently we, use a *step-down transformer*.

In a practical sense, we may describe the action as follows. The use of a step-down transformer allows the small resistance of R_L to appear as a much

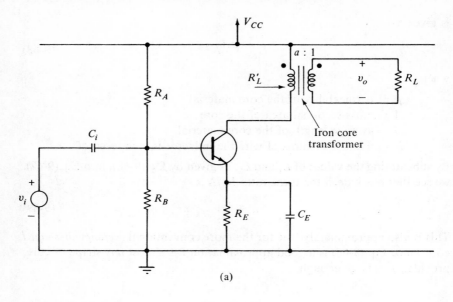

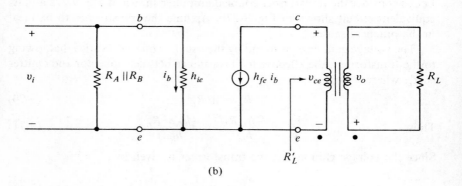

(a) A transformer coupled CE amplifier.
(b) The equivalent circuit of the transformer coupled CE amplifier.

FIGURE 9-27 A transformer coupled amplifier and equivalent circuit.

larger value to the collector of the transistor (namely R'_L, which is a^2 times as large as R_L). The quantity a is always larger than unity for a step-down transformer. The choice of a is left to the designer. Naturally, the value of a is chosen so that the quantity $a^2 R_L$ is about the same as the choice of a collector load resistor in an RC coupled amplifier. The quantity $a^2 R_L$ should be somewhat less than $1/h_{oe}$. Notice that the resulting gain in Eq. (9-73) for a transformer coupled amplifier with a step-down transformer is higher than that in Eq. (9-40) for an RC coupled amplifier having the same load.

It is interesting to perform a load line analysis for the transformer coupled amplifier shown in Fig. 9-27a. The resulting dc and ac load lines are shown in Fig. 9-28. Notice that the dc load line is almost vertical, since the transformer has no dc resistance and the slope of the dc load line is determined solely by R_E, which is usually quite small. The ac load line has a lesser slope, since the ac load resistance is now $a^2 R_L$.

Notice also that the collector-to-emitter peak-to-peak voltage v_{ce} may be larger than V_{CC}. This unusual phenomenon is explained by the energy-storing property of the inductors in the transformer.

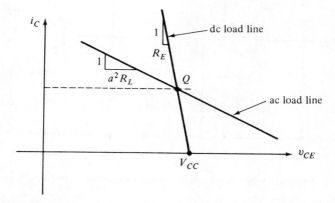

FIGURE 9-28 The ac and dc load lines of a transformer coupled amplifier such as that of Fig. 9-27(a).

9.10 Design of Transformer Coupled Amplifiers

Since a coupling transformer is comparatively bulky and expensive, its use in electronic circuits is decreasing as the price of transistors decreases. It is usually used for coupling very low impedance loads to a transistor amplifier, although, in may cases, an emitter follower is used. The designer may have to refer to economic considerations for his final choice in designing the circuit needed to meet a set of specifications.

There are two advantages in using a transformer coupled amplifier over an emitter follower: Larger voltage swings may be obtained, as illustrated by the load-line analysis in Section 9.9; and no additional energy is needed from the power supply. On the other hand, the disadvantages in using transformer coupling are numerous: Transformers are usually quite large and do not allow for miniaturization of circuits; the transformer needed may be expensive if a large frequency response is desired; the inductance of L_2 may have to be very large if a very low frequency response is desired; and large inductances tend to have large interwiring capacitance, which decreases the amplifier high-frequency response. In effect, this capacitance causes the output voltage to short-circuit at higher frequencies.

Transformer coupling may be used to couple two transistor stages, as shown in Fig. 9-29. In this type of coupling R_A and R_B are shorted out by

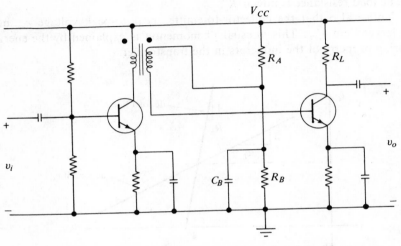

FIGURE 9-29 A two-stage transformer coupled amplifier.

C_B, increasing the overall gain. This leads only to a limited increase in gain and, therefore, is not used too frequently. Transformer coupled amplifiers, however, are used where amplifier efficiency is important, such as in power amplifiers. This type of design is considered again in Chapter 13, where power amplifiers are discussed in detail.

Example 9-5: Given a 10-V supply and an *npn* silicon transistor operating with an I_C of 4 mA. First design a transformer coupled amplifier to operate into an 8-Ω load with an overall voltage gain of about 20. Then obtain all the information necessary for purchasing the desired transformer, if the lowest frequency of operation of the amplifier is to be 20 Hz.

Solution:

The amplifier is biased in the usual manner using *H*-type biasing, so that I_C is 4 mA. The resulting circuit is shown in Fig. 9-30.

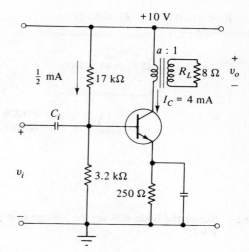

FIGURE 9-30 Amplifier circuit for Ex. 9-5.

$$r_e = m \times \frac{26 \text{ (V)}}{I_E(\text{mA})} = 1.4 \times \frac{26}{4} = 9.1 \, \Omega$$

assuming $m = 1.4$ for silicon.

To meet the gain condition we calculate a from the gain Eq. (9-73). The voltage gain is given by

$$A_v = \frac{aR_L}{r_e}$$

and a is

$$a = \frac{A_v r_e}{R_L} = \frac{20 \times 9.1}{8} \simeq 23$$

Now the required secondary inductance is calculated for the transformer:

$$L_2 > \frac{R_L}{\omega_L} = \frac{8}{2\pi \times 20} \simeq 64 \text{ mH}$$

Since the winding resistances are usually given by the manufacturer, we should try to obtain maximum allowable values. Let us assume that we are working within the usual accuracy of 10 per cent. The winding resistance R_2 of L_2 of the transformer should not be over 5 per cent of the load resistance R_L, and the winding resistance R_1 of L_1 should not be over 5 per cent of $a^2 R_L$. In this way, the voltage drop across these winding resistances does not lower the output voltage by more than 10 per cent; that is, R_1 should be less than $0.05 a^2 R_L$ and R_2 should be less than $0.05 R_L$:

$$R_1 < 210 \, \Omega \quad \text{and} \quad R_2 < 0.4 \, \Omega$$

The entire specification of the required transformer is summarized in Fig. 9-31.

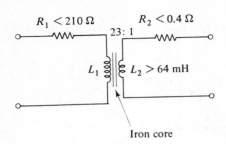

FIGURE 9-31 Circuit description of transformer required in solution for Ex. 9-5.

PROBLEMS

9.1 Sketch the Bode plot for the voltage gain v_o/v_i for the network shown in Fig. P 9-1.

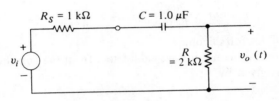

FIGURE P9-1

9.2 Increase C to 10.0 μF in Problem 9.1 and compare the resulting Bode plots.

9.3 Sketch the Bode plot for the voltage gain of the amplifier shown in Fig. 9-3a, if $R_S = 2$ kΩ, $C_i = 10\mu$F, $R_B = 1$ MΩ, $R_L = 2$ kΩ, and h_{ie} and h_{fe} for the transistor are 2 kΩ and 100, respectively.

9.4 If, for the circuit shown in Fig. 9-3, the lower cutoff frequency were specified to be not less than 10 Hz, calculate the required size of C_i for $R_S = 2$ kΩ, $R_B = 1$ MΩ, $R_L = 2$ kΩ, h_{ie} and h_{fe} for the transistor at 2 kΩ and 100, respectively.

9.5 (a) Determine the size of C_E that would be necessary in Fig. P 9-5 to obtain a lower cutoff frequency of 15 Hz. Assume C_i to be a large capacitor in your analysis. Assume T_1 to be a silicon transistor with a β of 100.
(b) What would a suitable value for C_i be so that the assumption in part (a) would be true?

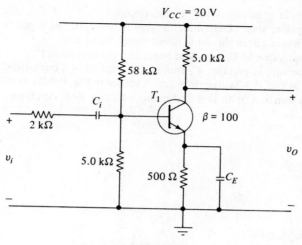

FIGURE P9-5

9.6 (a) Derive a design formula for obtaining the value of C_E when the lower cutoff frequency is specified for the circuit shown in Fig. P 9-6.
(b) Bias an amplifier of the type shown in Fig. P 9-6 to have a voltage gain v_o/v_i of -25 and lower cutoff frequency of approximately 20 Hz.

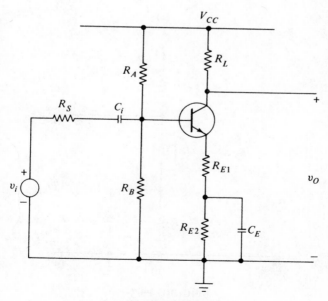

FIGURE P9-6

(c) Couple a second stage to the amplifier in Fig. P 9-6, using the same amplifier stage. Determine the size of the coupling capacitor needed to maintain about the same lower cutoff frequency.

(d) What value of C_i would be needed in the first stage?

9.7 Occasionally, in practice, C_i and R_g are omitted in a circuit such as the one shown in Fig. P 9-7a; the omission is shown in Fig. P 9-7b. In this case, both circuits function in an identical manner. Under what conditions is omission permissible?

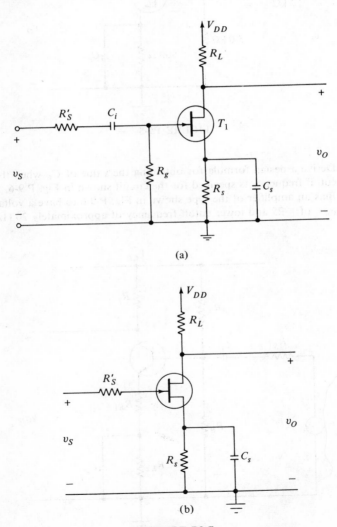

FIGURE P9-7

9.8 Determine the values of C_i and C_s for the circuit shown in Fig. P 9-7a, if $R'_S = 500 \text{ k}\Omega$, $R_g = 2 \text{ M}\Omega$, and $R_L = 5 \text{ k}\Omega$. The transistor T_1 has the following parameters: $g_m = 2.5 \text{ mA/V}$, $r_d = 80 \text{ k}\Omega$, and $I_{DSS} = 4.5 \text{ mA}$.

9.9 Obtain design formulas for C_B and C_i for the CB amplifier shown in Fig. P9-9.

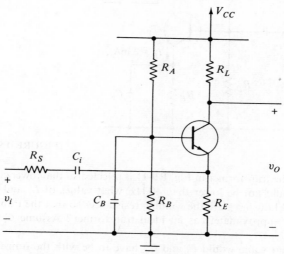

FIGURE P9-9

9.10 Obtain the values of C_i and C_B in Fig. P 9-9 if the following circuit values are given: $R_A = 17 \text{ k}\Omega$, $R_B = 3.2 \text{ k}\Omega$, $R_L = 4 \text{ k}\Omega$, $R_E = 1 \text{ k}\Omega$, $R_S = 35 \text{ }\Omega$, and $V_{CC} = 10 \text{ V}$. The transistor T_1 is a silicon *npn* transistor with $\beta = 100$ and $r_d = 60 \text{ k}\Omega$. The desired lower cutoff frequency is to be not less than 30 Hz.

9.11 (a) Assuming C_i and C_E to be large and the transformer to be ideal in Fig. P 9-11, sketch the dc and ac load lines for the transistor T_1.
(b) Determine the voltage gain v_o/v_i of the entire circuit.
(c) Calculate the maximum peak-to-peak voltage of v_{CE} and of v_o before clipping occurs.
(d) Calculate the maximum peak-to-peak input voltage v_i.

238 Chap. 9 / Low-Frequency Response of Transistor Amplifiers

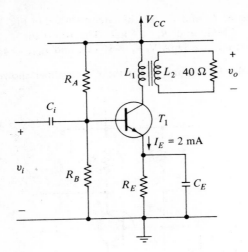

FIGURE P9-11

9.12 (a) If the transformer in Fig. P 9-11 is selected so that lower cutoff frequency should not be lower than 40 Hz, what values of L_1 and L_2 are needed and how large are the winding resistances so that the transformer would act approximately as an ideal transformer? Assume C_i and C_E to be large.

(b) What value would C_i and C_E have to be with the transformer selected in part (a) to still have the same frequency response? Assume the transistor to be silicon and have a β of 100.

9.13 Calculate the rms value of the voltage v_o in Problem 9.11c. Calculate the power delivered to the 40-Ω load with this voltage.

9.14 Design a transformer coupled amplifier to operate from a 10-V source and with an emitter current $I_E = 1$ mA in the transistor. The overall gain is to be 2.0 and load resistor is to be 4 Ω. Determine all necessary information about the transformer if lower cutoff frequency is to be 40 Hz.

REFERENCES

[1] SCHILLING, D. L., and BELOVE, C., *Electronic Circuits: Discrete and Integrated* (McGraw-Hill Book Co., New York, 1968).

[2] JOYCE, M. V., and CLARKE, K. K., *Transistor Circuit Analysis* (Addison-Wesley Publishing Co., Inc., Reading, Mass., 1961).

10

HIGH-FREQUENCY RESPONSE OF TRANSISTOR AMPLIFIERS

We have already seen that the lowest frequency at which an amplifier can operate is determined by the bypass and coupling capacitors in the circuit. These are elements that the designer has a good measure of control over. Generally speaking, the low-frequency response of an amplifier can be improved by using larger coupling and bypass capacitors.

The extent to which an amplifier can operate at high frequencies is not quite as simple. The high-frequency response is usually governed by the internal capacitances within the transistor, which we have omitted until now. Primarily, the models for the transistor were assumed to be resistive. At high frequencies, however, it is found that this simplification is no longer valid. In trying to improve the high-frequency response of an amplifier the designer usually has two alternatives: to choose a transistor with less internal capacitances, or to choose those configurations in which the frequency response is inherently quite large. These configurations are usually the common base, common gate, common collector, or common drain amplifiers. The common emitter and common source amplifiers find their use in the lower frequency range. They usually cover the audio frequencies quite well.

10.1 Hybrid-Pi Equivalent Circuit

The *hybrid-pi* equivalent circuit is probably the most widely accepted model for a transistor at high frequencies. It is shown in Fig. 10-1. The hybrid-pi model is accredited to L. J. Giacoletto, who first published it in the *RCA*

240   Chap. 10 / High-Frequency Response of Transistor Amplifiers

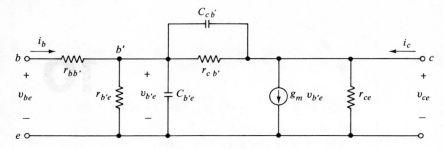

FIGURE 10-1   The complete hybrid-pi model for a transistor.

Review in December, 1954. The circuit was developed on the basis of the physical operation of the transistor, as well as on experimental work published in the same journal in June, 1953.

To begin our analysis and design work with the hybrid-pi model, we shall start at the low frequencies, where we have a fairly good understanding of the operation of a transistor from the previous sections. At low frequencies (generally through the audio range) the reactance of the capacitors $C_{cb'}$ and $C_{b'e}$ is large and can be neglected. Thus the low-frequency version of the hybrid-pi is the circuit given in Fig. 10-2. The resistor $r_{bb'}$ is the ohmic base resistance and is equal to the resistance r_b in the equivalent tee-model of a transistor.

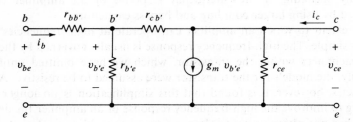

FIGURE 10-2   The low frequency version of the hybrid-pi model.

To relate the transistor model of Fig. 10-2 to the hybrid model (h-parameter model) we recognize that except for $r_{bb'}$ the circuit of Fig. 10-2 is the equivalent circuit obtained when the y-parameters for a two-port are used. Thus we consider the two-port network of Fig. 10-3a and the pair of equations that describe it to be

$$i_1 = y_{11}v_1 + y_{12}v_2 \qquad (10\text{-}1)$$
$$i_2 = y_{21}v_1 + y_{22}v_2 \qquad (10\text{-}2)$$

In matrix form, Eqs. (10-1) and (10-2) can be written as

$$\begin{bmatrix} i_1 \\ i_2 \end{bmatrix} = \begin{bmatrix} y_{21} & y_{12} \\ y_{21} & y_{22} \end{bmatrix} \begin{bmatrix} v_1 \\ v_2 \end{bmatrix} \qquad (10\text{-}3)$$

Sec. 10.1 / Hybrid-Pi Equivalent Circuit 241

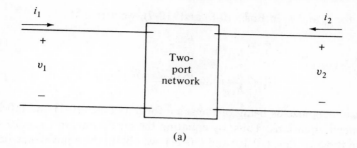

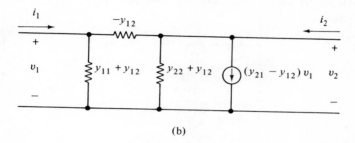

(a) A two-port network and
(b) its equivalent circuit using y parameters.

FIGURE 10-3 A two-port network.

The y-parameter equivalent circuit for a two-port is given in Fig. 10-3b, and the justification for this equivalent circuit is that it satisfies Eqs. (10-1) and (10-2). Notice that the form of the circuit of Fig. 10-3b is identical to Fig. 10-2 if $r_{bb'}$ is neglected or set equal to zero.

Next we consider the pair of equations that describes the two-port in terms of its h-parameters:

$$v_1 = h_{11}i_1 + h_{12}v_2 \tag{10-4}$$

$$i_2 = h_{21}i_1 + h_{22}v_2 \tag{10-5}$$

Equations (10-4) and (10-5) can be rewritten as

$$-h_{11}i_1 = -v_1 + h_{12}v_2 \tag{10-6}$$

$$-h_{21}i_1 + i_2 = h_{22}v_2 \tag{10-7}$$

Solving for i_1 in Eq. (10-6) and substituting this value into Eq. (10-7), Eqs. (10-6) and (10-7) become

$$h_{11}i_1 = -v_1 + h_{12}v_2 \tag{10-8}$$

$$-h_{21}\frac{-v_1 + h_{12}v_2}{-h_{11}} + i_2 = h_{22}v_2 \tag{10-9}$$

Solving for i_1 and i_2 in Eqs. (10-8) and (10-9), we arrive at

$$i_1 = \frac{1}{h_{11}}v_1 - \frac{h_{12}}{h_{11}}v_2 \tag{10-10}$$

$$i_2 = \frac{h_{21}}{h_{11}}v_1 + \frac{h_{11}h_{22} - h_{12}h_{21}}{h_{11}}v_2 \tag{10-11}$$

which we recognize as being of the same form as Eqs. (10-1) and (10-2), the *y*-parameter equations. Thus, by equating the coefficients in Eqs. (10-1) and (10-2) to those of Eqs. (10-10) and (10-11), we obtain the equivalence between the *h*- and *y*-parameters. Summarizing these equivalents, we obtain the following results:

$$y_{11} = \frac{1}{h_{11}} \tag{10-12}$$

$$y_{12} = \frac{-h_{12}}{h_{11}} \tag{10-13}$$

$$y_{21} = \frac{h_{21}}{h_{11}} \tag{10-14}$$

$$y_{22} = \frac{h_{11}h_{22} - h_{12}h_{21}}{h_{11}} \tag{10-15}$$

To establish the relationship between the hybrid-pi and the *h*-parameter equivalent circuit at low frequency, we consider each circuit with the ohmic base resistance removed, namely r_b and $r_{bb'}$, and assume that

$$r_b = r_{bb'} \tag{10-16}$$

These circuits are shown in Figs. 10-4a and 10-4b, respectively. Notice that the *y*-parameters and *h*-parameters of the resulting network have been indicated by the boxed-in symbols. Using the equivalents established by Eqs. (10-12) through (10-15), and these two circuits, we now obtain

$$r_{b'e} = \frac{1}{y_{11} + y_{12}} = \frac{h_{11}}{1 - h_{12}}$$

$$= \frac{h_{ie} - r_{bb'}}{1 - h_{re}} \tag{10-17}$$

$$\simeq h_{ie} - r_b$$

$$r_{cb'} = -\frac{1}{y_{12}} = \frac{h_{ie} - r_{bb'}}{h_{re}} \tag{10-18}$$

$$g_m = y_{21} - y_{12} = \frac{h_{21}}{h_{11}} + \frac{h_{12}}{h_{11}} = \frac{h_{fe} + h_{re}}{h_{ie} - r_{bb'}}$$

$$\simeq \frac{h_{fe}}{h_{ie} - r_{bb'}} \tag{10-19}$$

Sec. 10.1 / Hybrid-Pi Equivalent Circuit   243

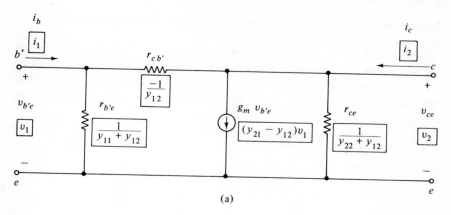

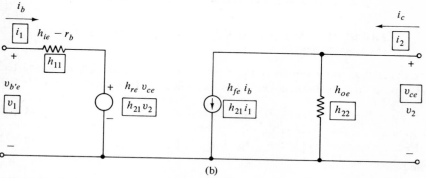

(a) A circuit showing a relationship between the y parameters and hybrid-pi parameters with r_b removed.

(b) Relationship between the h parameters of a transistor and the two-port h parameters with r_b removed from the circuit.

FIGURE 10-4 Relationship between various two-port and transistor parameters.

$$\frac{1}{r_{ce}} = y_{22} + y_{21} = \frac{h_{11}h_{22} - h_{12}h_{21}}{h_{11}} - \frac{h_{12}}{h_{11}}$$

$$= h_{22} - \frac{h_{12}(h_{21} + 1)}{h_{11}} \quad (10\text{-}20)$$

$$= h_{oe} - \frac{h_{re}(h_{fe} + 1)}{h_{ie} - r_{bb'}}$$

Equations (10-17) through (10-20) describe the resistors in the hybrid-pi

244 Chap. 10 / High-Frequency Response of Transistor Amplifiers

in terms of the low-frequency h-parameters* and the ohmic base resistance $r_{bb'}$, which we have previously studied.

An examination of these equations quickly reveals the fact that the resistor $r_{cb'}$ is quite large and in the order of several megohms. The usual procedure is to neglect it because of its large size and reduce the circuit of Fig. 10-2 to that given in Fig. 10-5. This approximate circuit is immediately

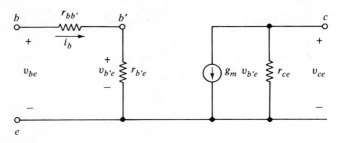

FIGURE 10-5 An approximate (or simplified) low frequency version of the hybrid-pi model.

recognized as being similar to the usual hybrid circuit when $h_{re}v_{ce}$ is omitted. By doing this we obtain a new set of equivalences. By inspection, we see that

$$r_{ce} = \frac{1}{h_{oe}} \simeq r_d \tag{10-21}$$

Also,
$$r_{bb'} + r_{b'e} = h_{ie} \tag{10-22}$$

and since
$$r_{bb'} = r_b$$
$$r_{b'e} = (\beta + 1)r_e \tag{10-23}$$

Now
$$g_m v_{b'e} = h_{fe} i_b \tag{10-24}$$

or
$$g_m = \frac{h_{fe} i_b}{v_{b'e}} = \frac{h_{fe} i_b}{r_{b'e} i_b}$$

$$= \frac{\beta}{(\beta + 1)r_e} \tag{10-25}$$

$$= \frac{\alpha}{r_e}$$

$$\simeq \frac{1}{r_e}$$

Summarizing, we can say that the effect of neglecting $r_{cb'}$ in the hybrid-pi model is the same as neglecting h_{re}, since it leads to a similar equivalent circuit. This circuit is shown in Fig. 10-6, with the r-parameter equivalents substituted for the circuit values of the hybrid-pi model.

---
* In Section 3.5 we assumed h-parameters to be simple resistors and ratios. The h-parameter concept may be extended to include a high frequency model for transistors.

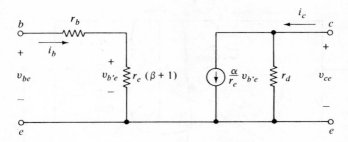

FIGURE 10-6 Approximate low frequency hybrid-pi model with r parameters used to designate the circuit.

From previous work with the h-parameter circuit, we realize that the circuit of Fig. 10-6 will describe the action of the transistor at low frequencies very well. The approximation made in neglecting h_{re} led to very little error in computing the circuit performance of the transistor amplifiers.

At the higher frequencies the circuit of Fig. 10-7 describes the action of

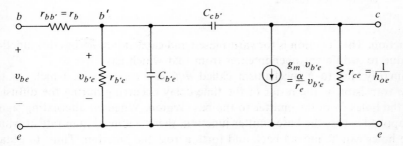

FIGURE 10-7 The approximate hybrid-pi model with the addition of the internal capacitors.

the transistor quite adequately. It is the hybrid-pi model with $r_{cb'}$ neglected. Physically, the capacitance $C_{cb'}$ represents the capacitance between the collector and the internal base region. Since the collector base junction is reverse biased, the extrinsic holes and electrons have moved away from the junction. In a *pnp* transistor, the result of the reverse bias is to move the electrons in the base region away from the junction and the holes in the collector away from the junction as well (Fig. 10-8). The result is that in the immediate vicinity of the base collector junction, the base has a positive charge and the collector a negative charge, similar to that of a capacitor. The amount of charge varies with the amount of reverse bias V_{CB}, and hence the capacitance $C_{cb'}$. Typical values of $C_{cb'}$ vary from 0.8 to about 50 pF. Of course, very high frequency transistors have the lowest value of $C_{cb'}$ possible.

The capacitance $C_{b'e}$ represents the capacitance across the emitter base

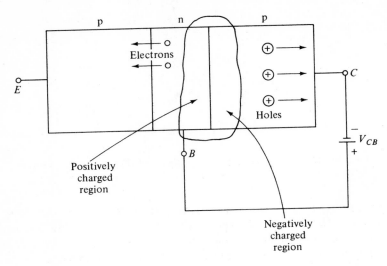

FIGURE 10-8 Action of holes and electrons in the reverse biased collector base region.

junction. This junction is forward-biased and capacitance across this junction is due to a different phenomenon from that which gave rise to $C_{cb'}$. $C_{b'e}$ is primarily due to a mechanism called *diffusion capacitance*, which, in the *pnp* transistor, is the result of the time delay occurring during the diffusion of the holes from the emitter to the base region. When an alternating signal is applied across the base emitter junction, there is a maximum rate at which the holes can be moved back and forth across the junction. Thus, to obtain a high-frequency transistor, the time delay should be short in comparison to the operating frequency. Typical values of $C_{b'e}$ range from about 20 pF to 0.01 μF. Typically, $C_{b'e}$ is about 100 times larger than $C_{cb'}$ but in general is a function of the dc emitter current I_E. The larger I_E, the larger the amount of charges moving across the base emitter junction and the larger the value of $C_{b'e}$.

The values of $C_{cb'}$ and $C_{b'e}$ can usually be obtained from the transistor specifications supplied by the manufacturer. $C_{b'c}$ is sometimes called C_{ob} and is usually plotted as a function of the reverse-biased voltage V_{CB}. Occasionally, the open-circuit output capacitance C_{obo} is given. This is the collector-to-emitter capacitance measured on a bridge with the base open-circuited. This value very closely approximates $C_{cb'}$, as will be seen in Section 10.3. The value of $C_{b'e}$ is obtained from a knowledge of f_T, the *gain–bandwidth product* of the transistor. The quantity f_T is usually plotted as a function of V_{CE} and I_C. A discussion of f_T and the relationship of f_T to $C_{b'e}$ will be covered in Section 10.2.

10.2 β Cutoff, α Cutoff, and Gain–Bandwidth Product

A complex β can be defined as the short-circuit current gain of a CE transistor at all frequencies. Symbolically, this can be written as

$$\beta(j\omega) = \frac{I_c}{I_b}\bigg|_{V_{ce}=0} \tag{10-26}$$

Using the hybrid-pi circuit of Fig. 10-7, shorting the output and driving the input with a current source of I_b, as shown in Fig. 10-9, we can arrive at

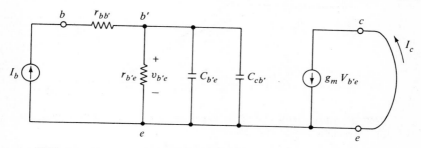

FIGURE 10-9 Circuit illustrating method for obtaining complex β: $\beta(j\omega)$ as function of frequency.

a value for I_c and hence a value for the complex β:

$$I_c\big|_{V_{ce}=0} = g_m V_{b'e} \tag{10-27}$$

but

$$V_{b'e}\big|_{V_{ce}=0} = I_b \frac{r_{b'e}}{1 + j\omega(C_{b'e} + C_{cb'})r_{b'e}} \tag{10-28}$$

and

$$\beta(j\omega) = \frac{g_m r_{b'e}}{1 + j\omega(C_{b'e} + C_{cb'})r_{b'e}} \tag{10-29}$$

Usually Eq. (10-29) is written as

$$\beta(j\omega) = \frac{\beta_o}{1 + \dfrac{j\omega}{\omega_\beta}} \tag{10-30}$$

where

$$\beta_o = g_m r_{b'e} \simeq h_{fe}$$

and

$$\omega_\beta = \frac{1}{r_{b'e}(C_{b'e} + C_{cb'})}$$

The quantity β_o is known as the low-frequency beta and the same beta that we have used previously, except that we were not concerned about current gain at high frequencies. Of course, β_o is approximately the same as h_{fe}. The frequency ω_β is known as the *beta (β) cutoff frequency* and is that value of ω when the magnitude of β is 3 dB below the low-frequency value β_o. A Bode plot of Eq. (10-30), and hence a plot of β as a function of frequency, is shown in Fig. 10-10. The frequency ω_T is usually called the *gain-bandwidth product of the transistor* and is the frequency at which the magnitude of β falls off to unity. From the properties of log-log plots it is easy to see that ω_T is given by

$$\omega_T = \beta_o \omega_\beta \qquad (10\text{-}31)$$

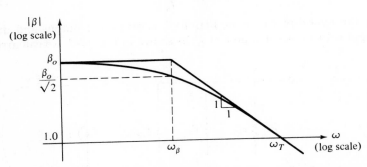

FIGURE 10-10 Bode plot of the magnitude of β illustrating the manner of determining β_o, ω_β, and ω_T.

The expressions for the β-cutoff and the gain-bandwidth product are also given by

$$f_\beta = \frac{1}{2\pi r_{b'e}(C_{b'e} + C_{cb'})} \qquad (10\text{-}32)$$

$$f_T = \beta_o f_\beta \qquad (10\text{-}33)$$

Since $r_{b'e}$ is equal to $(\beta_o + 1)r_e$, the substitution of Eq. (10-32) into Eq. (10-33) yields

$$f_T = \frac{\beta_o}{2\pi(\beta_o + 1)r_e(C_{b'e} + C_{cb'})} \qquad (10\text{-}34)$$
$$\simeq \frac{1}{2\pi r_e C_{b'e}}$$

if
$$C_{b'e} \gg C_{cb'}$$

Equation (10-34) can be used to obtain the value for $C_{b'e}$ in the hybrid-pi model:

$$C_{b'e} \simeq \frac{1}{2\pi r_e f_T} \qquad (10\text{-}35)$$

As mentioned earlier, the approximation made in Eqs. (10-34) and (10-35) is quite reasonable, since, in most cases, $C_{cb'}$ is about 100 times smaller than $C_{b'e}$. It is interesting to observe the dependence of $C_{b'e}$ on the dc emitter current I_E, since r_e, in Eq. (10-35), is a function of the emitter current.

A sketch of a typical plot of f_T as a function of V_{CE} and I_C is shown in Fig. 10-11.

Similarly, a complex α can now be defined as the short-circuit current gain of a CB transistor for all frequencies:

$$\alpha(j\omega) = \left.\frac{I_c}{I_e}\right|_{V_{cb}=0} \tag{10-36}$$

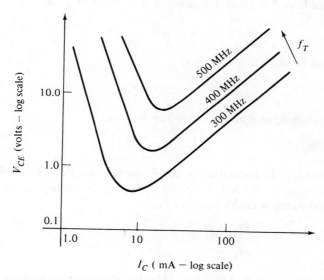

FIGURE 10-11 A sketch illustrating the contours of constant gain bandwidth product f_T.

The short-circuit current gain becomes

$$\alpha(j\omega) = \frac{\alpha_o}{1 + j\dfrac{\omega}{\omega_\alpha}} \tag{10-37}$$

where α_o is the low-frequency current gain of the transistor in the CB configuration and ω_α is the *alpha (α) cutoff frequency*, or the frequency at which α has dropped 3 dB below α_o.

Returning to our expression for the complex β,

$$\beta = \frac{\beta_o}{1 + j\dfrac{\omega}{\omega_\beta}} \tag{10-38}$$

We solve for the value of β at ω_T, remembering that ω_T is much larger than

ω_β and thus, from Eq. (10-38), we obtain

$$\beta\bigg|_{\omega=\omega_T} = \frac{\beta_o}{1+j\frac{\omega_T}{\omega_\beta}} \simeq \frac{\beta_o}{j\frac{\omega_T}{\omega_\beta}}$$

$$\simeq -j\frac{\omega_\beta \omega_o}{\omega_T} = -j \qquad (10\text{-}39)$$

Recalling from Eq. (8-2), the value of α in terms of β as

$$\alpha = \frac{\beta}{\beta+1} \qquad (10\text{-}40)$$

we can solve for the value of α at ω_T,

$$\alpha\bigg|_{\omega=\omega_T} \simeq \frac{-j}{-j+1} \qquad (10\text{-}41)$$

and the magnitude of α at this frequency becomes

$$\left|\alpha\right|_{\omega=\omega_T} = \frac{1}{\sqrt{2}} \qquad (10\text{-}42)$$

Equation (10-42) illustrates that at the frequency ω_T, the magnitude of α is approximately at its 3-dB-down point. Thus the frequency ω_T is approximately equal to the α-cutoff frequency ω_α.

$$\omega_\alpha \simeq \omega_T \qquad (10\text{-}43)$$

or $$f_\alpha \simeq f_T \qquad \text{where } f_\alpha = \frac{\omega_\alpha}{2\pi} \qquad (10\text{-}44)$$

Experimentally, we find that Eqs. (10-33) and (10-44) are in error. Usually, the α-cutoff is 20 to 100 per cent higher than the gain–bandwidth product f_T.

Example 10-1: The following data for a *pnp* silicon transistor biased at $V_{CE} = -5$ V were obtained from a manufacturer's specification sheet:

$$I_C = -10 \text{ mA}$$
$$f_T = 400 \text{ MHz}$$
$$h_{fe} = 100$$
$$\frac{1}{h_{oe}} = 30 \text{ k}\Omega$$
$$C_{ob} = 2 \text{ pF}$$

Obtain all the circuit parameters for the hybrid-pi model.

Solution:

$$r_{b'e} = (\beta_o + 1)r_e \simeq h_{fe}\frac{mv_T}{I_E}$$
$$\simeq (100)(3.5) = 350 \text{ }\Omega$$

$$C_{b'e} = \frac{1}{2\pi r_e f_T}$$
$$= \frac{1}{2\pi \times 3.5 \times 4 \times 10^8}$$
$$= 110 \text{ pF}$$
$$C_{cb'} = C_{ob} = 2 \text{ pF}$$
$$g_m = \frac{\alpha}{r_e} \simeq \frac{1}{r_e} = 28 \text{ mmhos}$$
$$r_{ce} \simeq \frac{1}{h_{oe}} = 30 \text{ k}\Omega$$

and, since $r_{bb'}$ is usually not given, we generally assume it to be 20 Ω for a silicon transistor, that is,

$$r_{bb'} = r_b = 20 \text{ }\Omega$$

10.3 Miller Effect

Consider the circuit shown in Fig. 10-12, consisting of a voltage source V_g, the source resistance R_G, the amplifier with a voltage gain of $A_v = V_o/V_a$, and the feedback element consisting of an admittance Y_f connected from the output to the input of the amplifier, as shown. The input admittance of the amplifier is given by

$$Y_a = \frac{I_a}{V_a} \qquad (10\text{-}45)$$

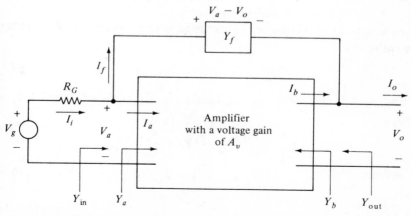

FIGURE 10-12 An amplifier with a feedback admittance Y_f.

and the output admittance by

$$Y_b = \frac{V_o}{-I_b} \qquad (10\text{-}46)$$

The input admittance of the amplifier and Y_f is designated by Y_{in}, given as

$$Y_{in} = \frac{I_i}{V_a} \qquad (10\text{-}47)$$

and the output admittance of the entire circuit Y_{out} is given as

$$Y_{out} = \frac{-I_o}{V_o} \qquad (10\text{-}48)$$

The input admittance Y_{in} can be expressed as

$$Y_{in} = \frac{I_i}{V_a} = \frac{I_a + I_f}{V_a} \qquad (10\text{-}49)$$

The current I_f can be expressed as

$$I_f = (V_a - V_o)Y_f \qquad (10\text{-}50)$$

The substitution of Eq. (10-50) into Eq. (10-49) yields

$$\begin{aligned} Y_{in} &= \frac{I_a}{V_a} + \frac{V_a - V_o}{V_a} Y_f \\ &= Y_a + (1 - A_v) Y_f \end{aligned} \qquad (10\text{-}51)$$

Equation (10-51) can now be interpreted as follows: The overall input admittance of the amplifier and the feedback element is the sum of the input admittance and feedback admittance, multiplied by 1, minus the voltage gain of the amplifier. If the gain of the amplifier itself is negative and real,

$$A_v = -G \qquad (10\text{-}52)$$

then the overall admittance is increased and the effect of Y_f, as seen by the input, is multiplied by the constant $G + 1$.

Similarly, at the output, we can write an expression for the admittance of the entire circuit:

$$Y_{out} = -\frac{I_o}{V_o} = -\frac{I_f + I_b}{V_o} \qquad (10\text{-}53)$$

Again, by substituting the expression for I_f, as given in Eq. 10-50, we obtain

$$\begin{aligned} Y_{out} &= -\left(\frac{V_a - V_o}{V_o} Y_f + \frac{I_b}{V_o}\right) \\ &= Y_f\left(1 - \frac{1}{A_v}\right) + Y_b \\ &\simeq Y_f + Y_b \quad \text{if } |A_v| \gg 1 \end{aligned} \qquad (10\text{-}54)$$

The interpretation for Eq. (10-54) is as follows: The overall output admittance is the sum of the output admittance of the amplifier and the feedback admittance, if the magnitude of the gain is substantially larger than unity. This is certainly the case for most transistor amplifiers (except for the emitter and source follower).

The results obtained in Eqs. (10-51) and (10-54) allow us to simplify the circuit of Fig. 10-12 to that of Fig. 10-13. Since Y_{in} and Y_{out} and V_o/V_a shown

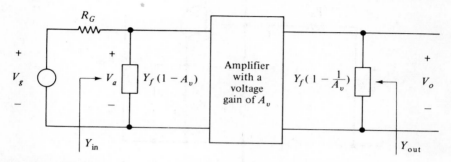

FIGURE 10-13 An equivalent circuit for an amplifier with feedback admittance Y_f as in Fig. 10-12.

in Fig. 10-13 are identical to that of Fig. 10-12, we conclude that the two circuits act identically as far as the voltage gain V_o/V_a is concerned. This transformation of Y_f to the input and output simplifies the analysis of the CE amplifier.

Consider the typical CE amplifier shown in Fig. 10-14a and its equivalent circuit, using the hybrid-pi model in Fig. 10-14b. To simplify the circuit of Fig. 10-14b, the Thévenin equivalent is taken to the left of the terminals b and e, and R_L and R_{ext} are combined into a single resistor R'_L, as shown in Fig. 10-15. In Fig. 10-15a we have replaced v_i, R_S, R_A, and R_B with v_g and R_G, whose values are given by

$$v_g = \frac{R_A \| R_B}{R_S + R_A \| R_B} v_i \tag{10-55}$$

and

$$R_G = R_A \| R_B \| R_S \tag{10-56}$$

In Fig. 10-15a, a box has been appropriately placed so that the circuit resembles that of Fig. 10-12. The boxed-in circuit is an amplifier with a voltage gain given as

$$A_v = \frac{V_o}{V_{b'e}} = -g_m R'_L \tag{10-57}$$

and the feedback admittance Y_f, given as

$$Y_f = j\omega C_{cb'} \tag{10-58}$$

By performing the same transformation as was previously described, the circuit in Fig. 10-15a reduces to that shown in Fig. 10-15b.

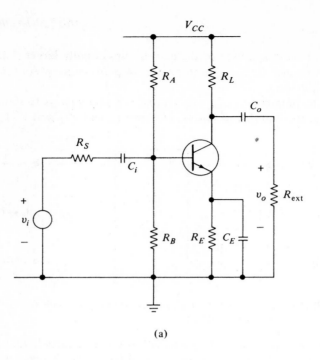

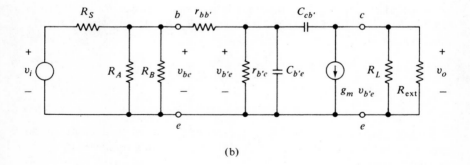

(a) A typical *CE* amplifier.
(b) The high frequency equivalent circuit using the hybrid-pi model.

FIGURE 10-14 The CE amplifier with an external load and its equivalent circuit.

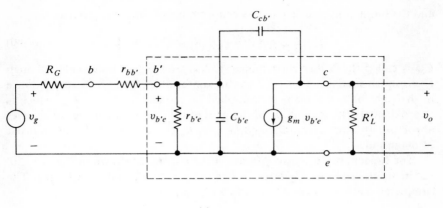

(a)

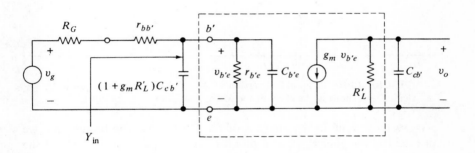

(b)

(a) Equivalent circuit of CE amplifier.

(b) A simplified version of Fig. 10–15a using a simple transformation of $C_{cb'}$ from a feedback element to an equivalent input and output element.

FIGURE 10-15 High frequency equivalent circuit of the CE amplifier.

The total admittance to the right of the terminals $b'e$ in Fig. 10-15a, as given by Eq. (10-51), yields

$$Y_{in} = Y_a + (1 - A_v)Y_f$$
$$= \frac{1}{r_{b'e}} + j\omega C_{b'e} + (1 + g_m R'_L)j\omega C_{cb'} \qquad (10\text{-}59)$$
$$= \frac{1}{r_{b'e}} + j\omega[C_{b'e} + (1 + g_m R'_L)C_{cb'}]$$

and the input capacitance C_M between terminals b' and e is given as

$$C_M = C_{b'e} + (1 + g_m R'_L)C_{cb'} = C_{b'e} + C'_M \qquad (10\text{-}60)$$

C'_M is called the Miller capacitance. It is interesting to note that, although $C_{cb'}$ is much smaller than $C_{b'e}$, it has an appreciable effect on the capacitance to the right of $b'e$, since it is multiplied by the factor $(1 + g_m R'_L)$, which is approximately the magnitude of the gain of the amplifier for the frequencies above ω_L. In most cases, $C_{b'e}$ and $(1 + g_m R'_L)C_{b'e}$ are about the same order of magnitude.

The capacitance $C_{cb'}$ also appears at the output, as shown in Fig. 10-15b, but its reactance is usually large in comparison to R'_L and is neglected. The completed equivalent circuit is shown in Fig. 10-16.

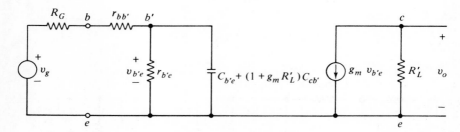

FIGURE 10-16 A simplified version of the CE high frequency equivalent circuit showing the effect of the Miller capacitance on the input.

As we have done in the analysis of the CE for the low frequencies, we shall calculate the gain V_o/V_g and plot it on a Bode plot. Since V_g is not the actual voltage gain but proportional to it, it will be understood that the actual gain can be obtained by using Eq. (10-55) to obtain V_o/V_i. The procedure is straightforward: the voltage $V_{b'e}$ is found by voltage division to be

$$V_{b'e} = \frac{r_{b'e}\dfrac{1}{sC_M} \bigg/ \left(\dfrac{1}{sC_M} + r_{b'e}\right)}{R_G + r_{bb'} + \dfrac{r_{b'e}}{1 + sC_M r_{b'e}}} V_g$$

$$= \frac{r_{b'e}}{r_{b'e} + (1 + sC_M r_{b'e})(R_G + r_{bb'})} V_g \qquad (10\text{-}61)$$

$$= \frac{\dfrac{r_{b'e}}{r_{b'e} + R_G + r_{bb'}} V_g}{1 + sC_M \dfrac{r_{b'e}(R_G + r_{bb'})}{r_{b'e} + R_G + r_{bb'}}}$$

Since the output voltage V_o is given by

$$V_o = -g_m R_L V_{b'e} \qquad (10\text{-}62)$$

the voltage gain V_o/V_g is found to be

$$\frac{V_o}{V_g} = \frac{-g_m R_L \frac{r_{b'e}}{r_{b'e} + R_G + r_{bb'}}}{1 + sC_M[r_{b'e}||(r_{bb'} + R_G)]} \quad (10\text{-}63)$$

The pertinent results of Eq. (10-63) can be quickly sketched by means of a Bode diagram, as shown in Fig. 10-17. The frequency ω_H is called the upper

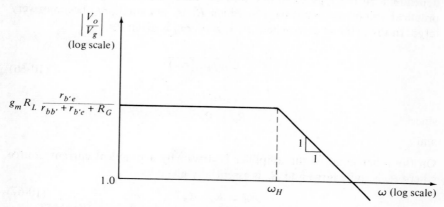

FIGURE 10-17 Bode diagram of Eq. 10-63 illustrating the high frequency response of a CE amplifier.

cutoff frequency of the amplifier and is the frequency at which the gain has decreased to 3 dB below the mid-band gain value. The upper cutoff frequency is given by

$$\omega_H = \frac{1}{C_M r_{b'e}||(R_G + r_{bb'})} \quad (10\text{-}64)$$

The expression

$$-g_m R_L \frac{r_{b'e}}{r_{b'e} + R_G + r_{bb'}}$$

which is proportional to the mid-band gain, can be shown to yield the same results for the frequencies below ω_H as were obtained in Chapter 4, where the coupling capacitors were considered to be short circuits, and a low-frequency model for the transistor was used:

$$\begin{aligned}
\frac{V_o}{V_g} &= -g_m R'_L \frac{r_{b'e}}{r_{b'e} + R_G + r_{bb'}} \\
&= \frac{-\alpha}{r_e} \frac{(\beta+1)r_e R'_L}{R_G + (\beta+1)r_e + r_{bb'}} \\
&= \frac{-\beta R'_L}{R_G + (\beta+1)r_e + r_{bb'}} \\
&\simeq -\frac{h_{fe} R'_L}{R_G + h_{ie}}
\end{aligned} \quad (10\text{-}65)$$

where
$$R_G = R_A || R_B || R_S$$

Equation (10-65) is precisely the same as Eq. (4-33). We have only replaced R_s by R_G, since R_G includes the effect of both the biasing resistors and the source resistance R_S in this analysis.

There are several important observations that can be made upon a closer examination of Eq. (10-64). First, the upper cutoff frequency ω_H is very dependent on the type of source that the amplifier is being driven by. If a practical voltage source is used in which R_S is very small, ω_H becomes very large. In the extreme case, where R_S is zero, ω_H is given by

$$\omega_H = \frac{1}{C_M(r_{b'e} || r_{bb'})} \quad (10\text{-}66)$$
$$\simeq \frac{1}{C_M r_{bb'}}$$

since
$$R_G = 0$$
and
$$r_{bb'} < r_{b'e}$$

On the other hand, if the amplifier is driven by a practical current source where R_S is quite large and R_G is essentially given by

$$R_G \simeq R_A || R_B \quad (10\text{-}67)$$

where
$$R_S > R_A || R_B$$

then the upper cutoff frequency decreases. An increase or decrease in the gain of the amplifier also has an effect on ω_H. The larger the gain, the larger C_M and hence the lower ω_H, and, similarly, the lower the gain, the higher ω_H one obtains.

In conclusion, we can say that to obtain a CE amplifier with a very large ω_H, and hence large high frequency, the designer has the following needs:

1. A source with a low output impedance.
2. The effective load resistance R'_L of the amplifier should be as low as possible; alternatively, the gain of the amplifier must be low.
3. A transistor with a large f_T, since $C_{b'e}$ is given by

$$C_{b'e} = \frac{1}{2\pi r_e f_T} \quad (10\text{-}68)$$

A larger f_T ensures a small value of $C_{b'e}$ and thus a lower value of C_M. It is also apparent from Eq. (10-68) that $C_{b'e}$ can be decreased by increasing the value of r_e and hence decreasing I_E, although this is not always possible, since f_T is a function of I_E (Fig. 10-11). A decrease in I_E often leads to a decrease in f_T; the designer must try to pick the best Q-point based on both of these considerations.
4. A large collector-to-emitter voltage V_{CE}, since this tends to decrease $C_{b'e}$; also, this usually tends to increase f_T somewhat.

Example 10-2: The common emitter amplifier shown in Fig. 10-18 uses a transistor with following specified parameters:

$$f_T = 400 \text{ MHz}$$
$$C_{cb'} = 2 \text{ pF}$$
$$\beta_o = 100$$

Determine the upper cutoff frequency of the amplifier.

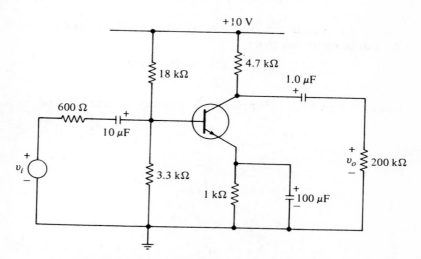

FIGURE 10-18 Circuit for example problem 10-2.

Solution:

First of all, the hybrid-pi parameters are established. From the circuit it is easy to establish the biasing current I_E to be

$$I_E \simeq 1.0 \text{ mA}$$

This gives a value for $r_{b'e}$:

$$r_{b'e} = (\beta_o + 1)r_e \simeq \frac{100 \times 1.4 \times 0.026}{1.0 \times 10^{-3}} = 3.5 \text{ k}\Omega$$

The capacitance $C_{b'e}$ can now be found from f_T:

$$C_{b'e} = \frac{1}{2\pi f_T r_e} = \frac{1}{2\pi \times 4 \times 10^8 \times 35} = 11.3 \text{ pF}$$

Since $r_{bb'}$ is not given, we shall assume it to be 20 Ω:

$$r_{bb'} \simeq 20 \text{ }\Omega$$

The value of $C_{cb'}$ is given as

$$C_{cb'} = 2 \text{ pF}$$

and since no value of r_{ce} is given, we shall assume it to be large in comparison to the load resistance.

The next step is to calculate the value of the capacitance C_M:

$$C_M = C_{b'e} + (1 + g_m R'_L)C_{cb'}$$
$$\simeq 11.3 + \left(1 + \frac{4.7 \times 10^3}{35}\right)2.0$$
$$= 11.3 + (135 \times 2.0)$$
$$= 281 \text{ pF}$$

The effect source resistance R_G as seen to the left of the base-emitter terminals, can be calculated to be

$$R_G = 600 \,\Omega \,||\, 18 \text{ k}\Omega \,||\, 3.3 \text{ k}\Omega$$
$$\simeq 490 \,\Omega$$

Using the expression for upper cutoff frequency, we finally obtain

$$\omega_H = \frac{1}{C_M[r_{b'e} \,||\, (R_G + r_{bb'})]}$$
$$= \frac{1}{2.81 \times 10^{-10} \times 4.4 \times 10^2}$$
$$= 8 \times 10^6 \text{ rad/s}$$

or
$$f_H = \frac{\omega_H}{2\pi} \simeq 1.3 \text{ MHz}$$

The value obtained in this example for f_H, the upper cutoff frequency, is typical of values usually obtained for common emitter amplifiers. A good check for f_H is to compare it with f_β. They are usually of the same order of magnitude, unless the amplifier is driven by a source with a very low source resistance. In practice, this is not usually the case.

10.4 Total Frequency Response of the CE Amplifier

To illustrate the complete frequency response of a transistor amplifier, we shall choose the current biased CE amplifier shown in Fig. 10-19a. This example is chosen because it contains only one external capacitance and easily illustrates the method for obtaining the complete frequency response of an amplifier. The H-type biased amplifier is left as an exercise.

The equivalent circuit of the amplifier in Fig. 10-19a is shown in Fig. 10-19b. The hybrid-pi model is used here and C_i is left in the circuit so that its effect can be analyzed. Since C_i is several thousand times larger than $C_{cb'}$ or $C_{b'e}$ at the low frequencies, $C_{cb'}$ and $C_{b'e}$ can be considered to be open circuits, while C_i may not be neglected. Therefore, the circuit Fig. 10-19b simplifies to that shown in Fig. 10-20.

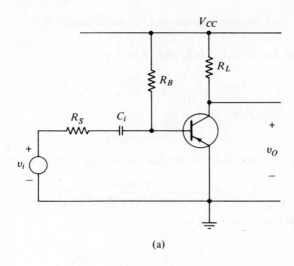

(a)

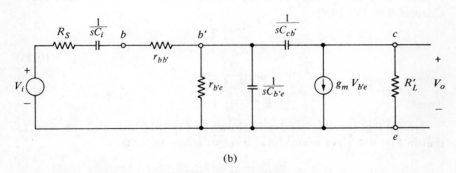

(b)

(a) A simple CE amplifier with an input coupling capacitor.
(b) The combined low- and high-frequency equivalent circuit.

FIGURE 10-19 The CE amplifier and its complete equivalent circuit for both high and low frequencies.

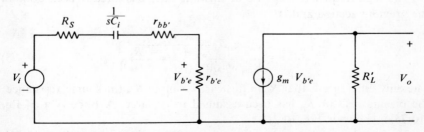

FIGURE 10-20 The low-frequency equivalent circuit of a CE amplifier using the hybrid-pi model.

Using voltage division, $V_{b'e}$ is found to be

$$V_{b'e} = \frac{sC_i r_{b'e}}{1 + sC_i(R_S + r_{bb'} + r_{b'e})} V_i \qquad (10\text{-}69)$$

and the output voltage V_o is given by

$$V_o(s) = -g_m V_{b'e} R'_L \qquad (10\text{-}70)$$

Combining Eqs. (10-69) and (10-70), the voltage gain $A_v = V_o/V_i$ is

$$A_v = \frac{V_o}{V_i} = \frac{-g_m R'_L sC_i r_{b'e}}{1 + sC_i(R_S + r_{bb'} + r_{b'e})} \qquad (10\text{-}71)$$

Equation (10-71) is of the form of

$$A_v = \frac{-g_m R'_L \dfrac{r_{be}}{R_S + r_{bb'} + r_{b'e}} \dfrac{s}{\omega_L}}{1 + (s/\omega_L)} \qquad (10\text{-}72)$$

A Bode plot of Eq. (10-72) is shown in Fig. 10-21a.

The value of ω_L obtained in Eq. (10-72) is identical to that obtained in Chapter 9 in Eq. (9-9):

$$\omega_L = \frac{1}{C_i(R_S + r_{bb'} + r_{b'e})}$$

$$= \frac{1}{C_i(R_S + h_{ie})} \qquad (10\text{-}73)$$

since
$$h_{ie} = r_{bb} + r_{b'e}$$

Notice that the mid-band gain shown in Fig. 10-21a is also the same as that in Fig. 9-4, since upon suitable substitutions we obtain

$$g_m R_L \frac{r_{b'e}}{R_S + r_{bb} + r_{b'e}} = \frac{(\alpha/r_e) R_L (\beta_o + 1) r_e}{R_S + h_{ie}} \qquad (10\text{-}74)$$

$$= \frac{h_{fe} R_L}{R_S + h_{ie}}$$

The high-frequency response of this amplifier has already been done in the previous section and is

$$A_v(s) = \frac{-g_m R_L \dfrac{r_{b'e}}{r_{b'e} + r_{bb'} + R_S}}{1 + sC_M[r_{b'e} || (R_S + r_{bb'})]} \qquad (10\text{-}75)$$

The only difference is that R_G in this case is simply R_S, the source resistance; the biasing resistor R_B has been assumed to be large. A Bode plot of Eq. (10-75) is shown in Fig. 10-21b.

The low-frequency and high-frequency effects of Figs. 10-21a and 10-21b can be combined into a single Bode plot, as shown in Fig. 10-22, and are valid for all frequencies. From Fig. 10-22 a simple expression can be written for

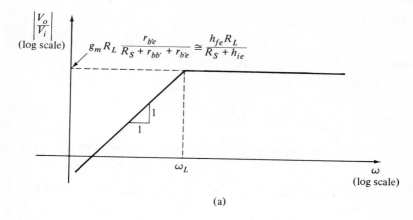

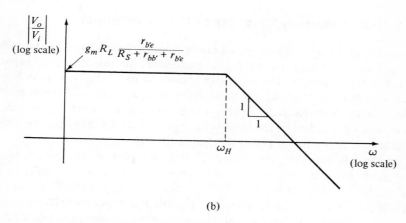

(a) Bode plot of the low-frequency response of a CE amplifier using the hybrid-pi model.
(b) The high-frequency response of the same amplifier.

FIGURE 10-21 Frequency response curves of a CE amplifier.

the voltage gain:

$$A_v(s) = \frac{-h_{fe}R_L}{h_{ie} + R_S} \frac{s/\omega_L}{(1 + s/\omega_L)(1 + s/\omega_H)} \quad (10\text{-}76)$$

where

$$\omega_L = \frac{1}{C_i(R_S + h_{ie})}$$

and

$$\omega_H = \frac{1}{C_M[r_{b'e} || (R_S + r_{bb'})]}$$

The frequencies ω_H and ω_L are, of course, the upper and lower cutoff frequencies of the amplifiers. At these frequencies the gain is 3 dB below the maximum gain that occurs in the band of frequencies between ω_H and ω_L.

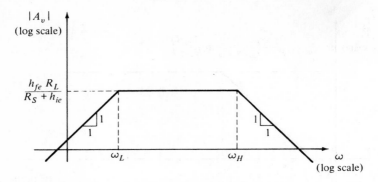

FIGURE 10-22 The combined low- and high-frequency Bode plot of a CE amplifier.

10.5 High-Frequency Response of the CB Amplifier

In the CE amplifier it was seen the upper frequency cutoff was in the order of f_β. This falloff in gain at the higher frequencies was due to internal capacitances in the transistor. The input capacitance in particular was quite large, because a portion of it was due to an internal capacitance multiplied by the gain. Even if transistors with very small capacitances were chosen, their effect, when multiplied by the large voltage gain of the CE amplifier, becomes substantial. This effect is called the Miller effect.

In the CB amplifier, the Miller effect is not present, since there is no substantial capacitance from the output to input (i.e., from the collector to the emitter). Furthermore, even if there were, the positive voltage gain of the CB amplifier would tend to reduce the effective input capacitance.

Let us consider the usual CB amplifier circuit as illustrated in Fig. 10-23a. To obtain the equivalent circuit for high frequencies for this amplifier, we first draw the hybrid-pi model for the transistor, as is shown in Fig. 10-23b (i.e., the portion enclosed in the curve). Then we connect R_L between collector and base, R_E between emitter and base, as well as R_S and v_i, observing the correct polarity of v_i. A ground symbol is placed next to the base terminal to indicate that it is common. In the equivalent circuit we have omitted r_{ce}, assuming it to be a large resistance.

The circuit of 10-23b is then redrawn into a more convenient form in Fig. 10-23c. The base, which is common, has been drawn at the bottom of the circuit, and v_i, R_S, and R_E have been replaced by a Thévenin equivalent consisting of v_g and R_G, where

$$v_g = \frac{R_E}{R_S + R_E} v_i \tag{10-77}$$

and
$$R_G = R_E \| R_S \tag{10-78}$$

Although the circuit of Fig. 10-23c may be analyzed with some difficulty

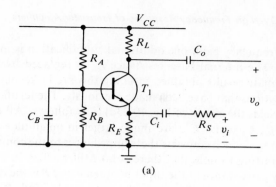

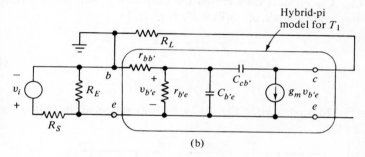

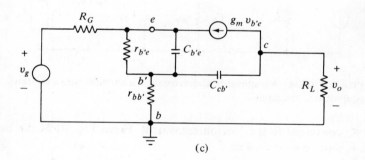

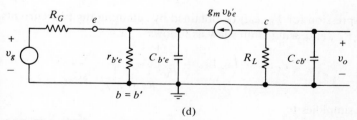

(a) A *CB* amplifier.
(b) The equivalent high-frequency circuit of the *CB* amplifier.
(c) A simplified high-frequency model
(d) Further simplification by removal of $r_{bb'}$.

FIGURE 10-23 Common base amplifier and equivalent circuits.

for its high-frequency gain, one additional simplification is made to ease the calculations. The internal base resistance $r_{bb'}$ is replaced by a short circuit. The approximate results obtained in doing this are in keeping with the type of approximations that have been made previously. The justification for doing this (which yields the circuit of Fig. 10-23d) is as follows: All the impedances offered by $r_{b'e}$, $C_{b'e}$, and $C_{cb'}$ are much larger in magnitude throughout the analysis than $r_{bb'}$, which connects these impedances to the common terminal b.

It is interesting to note that there is no Miller effect in Fig. 10-23d and that the only capacitance at the input between e and b is the diffusion capacitance $C_{b'e}$. $C_{cb'}$ appears only across the output. $C_{cb'}$ and R_L are combined into a single impedance, Z_L, where Z_L is given by

$$Z_L = R_L \| \frac{1}{sC_{cb'}} \tag{10-79}$$
$$= \frac{R_L}{1 + sC_{cb'}R_L}$$

Figure 10-23d is redrawn in Fig. 10-24, with Z_L shown at the output and with

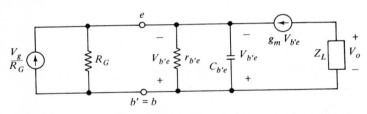

FIGURE 10-24 An equivalent circuit for the CB amplifier with a Norton equivalent at the input.

v_g and R_G converted to the Norton equivalent. From Fig. 10-24 the output voltage V_o can be expressed as

$$V_o = -g_m V_{b'e} Z_L \tag{10-80}$$

The expression for $V_{b'e}$ can be obtained by summing the two currents V_g/R_g and $g_m V_{b'e}$, and using Ohm's law as follows:

$$V_{b'e} = -\left(g_m V_{b'e} + \frac{V_g}{R_G}\right) \frac{1}{\frac{1}{R_G} + \frac{1}{r_{b'e}} + sC_{b'e}} \tag{10-81}$$

which simplifies to

$$V_{b'e} = -\left(g_m V_{b'e} + \frac{V_g}{R_G}\right) \frac{R_G r_{b'e}}{r_{b'e} + R_G + sR_G r_{b'e} C_{b'e}} \tag{10-82}$$

Collecting all the terms containing $V_{b'e}$ to the right-hand side and all other

Sec. 10.5 / High-Frequency Response of the CB Amplifier 267

terms to the left-hand side of the equation, we now obtain

$$V_{b'e}\left(1 + \frac{g_m R_G r_{b'e}}{r_{b'e} + R_G + sR_G r_{b'e} C_{b'e}}\right) = -V_g \frac{r_{b'e}}{r_{b'e} + R_G + sR_G r_{b'e} C_{b'e}} \tag{10-83}$$

Substitution of Eq. (10-80) into Eq. (10-83) allows us to calculate the voltage gain, V_o/V_g:

$$\frac{V_o}{V_g} = \frac{g_m Z_L r_{b'e}}{r_{b'e} + R_G + g_m r_{b'e} R_G + sR_G r_{b'e} C_{b'e}} \tag{10-84}$$

which simplifies to

$$\frac{V_o}{V_g} = \frac{g_m Z_L \frac{r_{b'e}}{r_{b'e} + (\beta_o + 1)R_G}}{1 + s\frac{R_G r_{b'e}}{r_{b'e} + (\beta_o + 1)R_G} C_{b'e}}$$

$$= \frac{g_m Z_L \frac{r_e}{r_e + R_G}}{1 + s\frac{R_G r_e}{r_e + R_G} C_{b'e}} \tag{10-85}$$

$$= \frac{\frac{Z_L}{r_e + R_G}}{1 + s\frac{R_G r_e}{r_e + R_G} C_{b'e}}$$

Substitution of the value for Z_L, as given in Eq. (10-79), into Eq. (10-85) yields

$$\frac{V_o}{V_g} = \frac{\frac{R_L}{r_e + R_G}}{\left(1 + s\frac{R_G r_e}{r_e + R_G} C_{b'e}\right)(1 + sC_{cb'} R_L)} \tag{10-86}$$

$$= \frac{\frac{R_L}{r_e + R_G}}{(1 + s/\omega_{H1})(1 + s/\omega_{H2})}$$

where
$$\omega_{H1} = \frac{1}{C_{b'e}(R_G || r_e)}$$

and
$$\omega_{H2} = \frac{1}{C_{cb'} R_L}$$

Equation (10-86) is the approximate frequency response for the CB amplifier. A Bode plot for this equation is shown in Fig. 10-25, although the relative values of ω_{H1} and ω_{H2} may not be in the order shown.

If the time constant $C_{cb'} R_L$ is larger than $C_{b'e}(R_G || r_e)$, then the 3-dB upper cutoff frequency ω_H is given approximately as

$$\omega_H \simeq \frac{1}{C_{cb'} R_L} \tag{10-87}$$

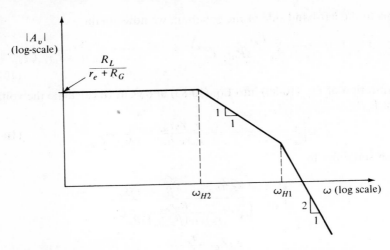

FIGURE 10-25 The Bode plot of the high-frequency response of a CB amplifier.

However, if the time constant $C_{cb'}R_L$ is smaller than $C_{b'e}(R_G||r_e)$, then the upper cutoff frequency becomes

$$\omega_H \simeq \frac{1}{C_{b'e}(R_G||r_e)} \tag{10-88}$$

If the value of ω_H, as given by Eq. (10-88), is compared with the gain-bandwidth product ω_T of a transistor,

$$\omega_T = \frac{1}{C_{b'e}r_e} \tag{10-89}$$

we see that the upper cutoff frequency is larger than ω_T. This is not usually the case in practice; the upper cutoff frequency is generally given by Eq. (10-87). However, this number is usually in the same order of magnitude as ω_T.

In conclusion we can therefore say that the frequency response of a CB amplifier is much better than that of a CE amplifier; however, the CB amplifier has several major drawbacks:

1. More than one stage of a CB amplifier cannot be cascaded together, since the second CB amplifier in a cascade will load down the first, reducing the voltage gain to approximately unity.
2. Depending on the load and source impedances, at high frequencies the CB amplifier can become unstable.
3. The input impedance of a CB amplifier is too low to be used with many sources whose output impedance can be quite high.

Example 10-3:* The common base amplifier shown in Fig. 10-26 uses a transistor with the following specified parameters:

$$f_T = 400 \text{ MHz}$$
$$C_{cb'} = 2 \text{pF}$$
$$\beta_o = 100$$

Determine the upper cutoff frequency of the amplifier.

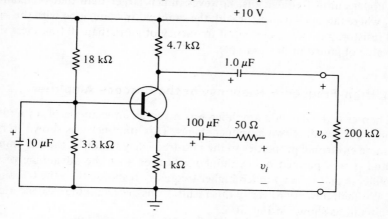

FIGURE 10-26 Circuit of example problem 10-3.

Solution:

Since this amplifier is biased at the same Q-point and has the same specifications as that of Example 10-2, the transistor parameters are the same and are listed here:

$$r_e = 35 \text{ }\Omega$$
$$C_{b'e} = 11.3 \text{ pF}$$
$$C_{c'b} = 2.0 \text{ pF}$$

The source impedance R_G is given by

$$R_G = 1 \text{ k}\Omega || 50 \text{ }\Omega \simeq 50 \text{ }\Omega$$

The upper break frequencies can now be calculated:

$$\omega_{H2} = \frac{1}{C_{cb'}R_L}$$
$$= \frac{1}{2.0 \times 10^{-12} \times 4.7 \times 10^3}$$
$$= 106 \text{ Mrad/s}$$
$$f_{H2} = 17 \text{ MHz}$$

*Note the similarity between the statement of Example 10-3 and that of Example 10-2.

Since f_{H1} is generally larger than f_T, we need not calculate it. The upper cutoff frequency in this example is given by f_{H2}:

$$f_H = f_{H2} = 17 \text{ MHz}$$

In this example the upper cutoff frequency is substantially lower than f_T because of the very large load resistor R_L. If the upper cutoff frequency were to be made large, we would have to decrease R_L and hence the mid-band gain. The upper cutoff frequency is, however, much larger than that of Example 10-2, where the same transistor with the same parameters, used with the same load resistor, gave an upper cutoff frequency of more than 10 times less than the value obtained in this example.

10.6 High-Frequency Response of the Cascode Amplifier

In Chapter 6 the cascode amplifier was given as an example of an unusual manner of cascading two amplifier stages. This amplifier was shown to have the same electronic properties in the mid-frequency region as the CE amplifier, and it was pointed out without verification that the advantage of this amplifier is that it had a much higher frequency response than the CE amplifier. We shall now try to verify this result. Consider again the cascode amplifier circuit, as shown in Fig. 10-27.

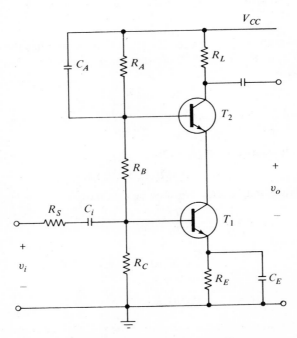

FIGURE 10-27 The cascode amplifier.

Sec. 10.6 / High-Frequency Response of the Cascode Amplifier

Recognizing the fact that the circuit of Fig. 10-27 is a CE–CB cascade, the equivalent circuit (Fig. 10-28a) is easily obtained. All the transistor parameters associated with transistor T_1 have been assigned an additional subscript, 1, and, similarly, a subscript 2 has been added for the parameters of transistor T_2.

To simplify the circuit of Fig. 10-28a to that of Fig. 10-28b, we apply the transformation of Section 10.3 to obtain the capacitance C_M at the input and the resulting output capacitance of the CE stage. The circuit can now be clearly divided into two parts, as indicated by the dashed line a-b. To the left of a-b we have essentially a CB amplifier. Thus we conclude that the frequency response of this portion of the circuit, for typical values of R_L, $C_{cb'2}$, and $C_{b'e2}$, is quite high.

The frequency response is therefore limited by the frequency response of the gain $V_{b'e1}/V_g$, which can be expressed by using voltage division as

$$\frac{V_{b'e1}}{V_g} = \frac{\dfrac{r_{b'e}}{1+sr_{b'e1}C_M}}{R_g + \dfrac{r_{b'e}}{1+sr_{b'e1}C_M}}$$

$$= \frac{r_{b'e1}}{R_g + r_{b'e1} + sC_M r_{b'e1} R_g} \qquad (10\text{-}90)$$

$$= \frac{\dfrac{r_{b'e1}}{R_g + r_{b'e1}}}{1 + s/\omega_H}$$

where
$$\omega_H = \frac{1}{C_M(r_{b'e} \| R_g)}$$
$$C_M = 2C_{cb1} + C_{be2}$$
and
$$R_g = (R_C \| R_B \| R_S) + r_{bb'}$$

It is interesting to note the size of the Miller capacitance C'_M in Eq. (10-90):

$$C_M = C_{b'e1} + (1 + g_{m1}R_{L1})C_{cb'1} = C_{be1} + C'_M \qquad (10\text{-}91)$$

where R_{L1} is the input impedance of the second stage. The impedance R_{L1} is the effective load of the CE amplifier and is essentially r_{e2} for the following reason: The CB amplifier has a much wider frequency response than the CE amplifier, and, therefore, at the frequencies at which the gain of CE is falling off, the CB low-frequency equivalent circuit can still be used:

$$R_{L1} = r_{e2} \qquad (10\text{-}92)$$

Since essentially the same dc emitter current flows in both T_1 and T_2, we conclude that

$$g_{m1} \simeq \frac{1}{r_{e2}} \qquad (10\text{-}93)$$

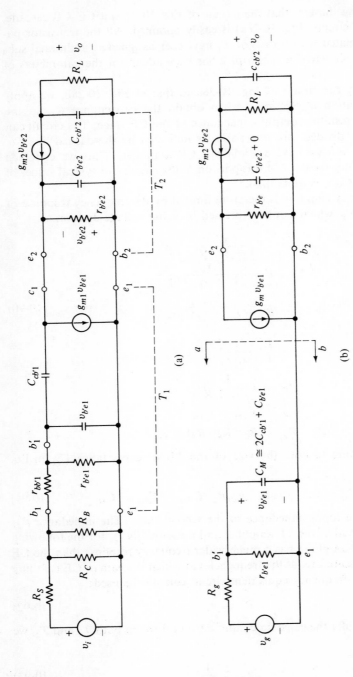

FIGURE 10-28 Cascode amplifier equivalent circuit.
(a) The complete high-frequency equivalent circuit of the cascode amplifier.
(b) A simplification of this circuit.

Thus C_M can be approximated as

$$C_M = C_{b'e1} + (1 + g_m R_{L1})C_{cb'}$$
$$\simeq C_{b'e1} + 2C_{cb'1} \qquad (10\text{-}94)$$

This value is much lower than for most CE amplifiers, because the gain of the first stage is only -1.0.

Thus, in calculating ω_H, which now becomes the upper cutoff frequency of the entire circuit, we obtain

$$\omega_H = \frac{1}{C_M(r_{b'e} \| R_g)} \qquad (10\text{-}95)$$

where
$$C_M = C_{b'e1} + 2C_{cb'1}$$

We see that ω_H is higher than that for a CE amplifier with R_L as the load, because the size of C_M has been reduced substantially. In practice, the improvement is about a factor of 2 or 3 depending upon the relative sizes of $C_{cb'}$, $C_{b'e}$, and R_g.

In conclusion, we note that the gain of the cascode amplifier is the same as that for a CE amplifier, but the frequency response has been increased, because the Miller capacitance has been reduced in the first stage. A closer look at ω_H in Eq. (10-95) indicates that this value can approach ω_T if R_g is small. In this case, the frequency response will be limited by the frequency response of the CB amplifier. If this is the case, a more complete analysis is necessary, in which the value of C_M in Eq. (10-95) must be recalculated.

10.7 Frequency Response of the Emitter Follower

The frequency response of the emitter follower is also much better than that of the CE amplifier. Consider the emitter follower shown in Fig. 10-29a. We shall assume that it is biased by a previous stage with an output resistance of R_S.

The high-frequency equivalent circuit is obtained by using the hybrid-pi model and is shown in Fig. 10-29b. To simplify the circuit, the Thévenin equivalent is taken to the left of the dotted line in Fig. 10-29b and redrawn in the transformed version shown in Fig. 10-29c, where $Z_{b'e}$ and Z_S are given by the expressions

$$Z_{b'e} = \frac{r_{b'e}}{1 + sC_{b'e}r_{b'e}} \qquad (10\text{-}96)$$

$$Z_S = \frac{R_S + r_{bb'}}{1 + sC_{cb'}(R_S + r_{bb'})}$$
$$\simeq \frac{R_S}{1 + sC_{cb'}R_S} \qquad \text{since } R_S > r_{bb'} \qquad (10\text{-}97)$$

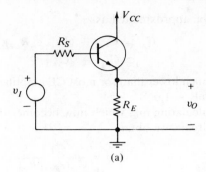

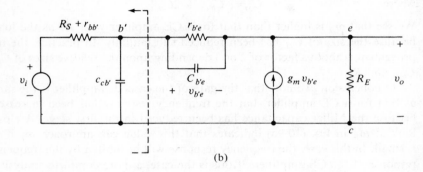

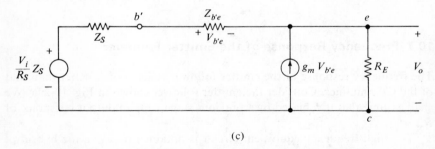

FIGURE 10-29 CC amplifier and equivalent circuit. (a) Biased emitter follower. (b) Equivalent circuit of the emitter follower. (c) The simplified equivalent circuit.

The ohmic base resistance is assumed to be negligible in Eq. (10-96) and in the expression for the Thévenin source in Fig. 10-29c.

The output voltage V_o can be obtained by summing the currents at node e in Fig. 10-29c:

$$\begin{aligned}V_o &= \left(g_m V_{b'e} + \frac{V_{b'e}}{Z_{b'e}}\right)R_E \\ &= V_{b'e}\left(g_m + \frac{1}{Z_{b'e}}\right)R_E \qquad (10\text{-}98) \\ &\simeq g_m V_{b'e} R_E \qquad \text{if } |Z_{b'e}| \gg r_e\end{aligned}$$

The approximate expression obtained in Eq. (10-98) is valid to magnitude or so below f_T, since it can easily be shown that the magnitude of $|Z_{b'e}|$ at f_T is approximately r_e. The assumption made to obtain in Eq. (10-98) should be kept in mind when the end results of this discussion are used.

From Eq. (10-98) we now write

$$V_{b'e} = \frac{V_o}{g_m R_E} \quad \text{for } \omega < f_T \tag{10-99}$$

By the use of Kirchhoff's voltage law in Fig. 10-29c, we write

$$\frac{V_i}{R_S} Z_S = (Z_S + Z_{b'e}) \frac{V_{b'e}}{Z_{b'e}} + V_o \tag{10-100}$$

Substituting Eq. (10-99) into Eq. (10-100), we obtain

$$\frac{V_i}{R_S} Z_S = \left[(Z_S + Z_{b'e}) \frac{1}{Z_{b'e} g_m R_E} + 1 \right] V_o \tag{10-101}$$

Solving for the reciprocal of the voltage gain, we now obtain

$$\begin{aligned}\frac{1}{A_v} = \frac{V_i}{V_o} &= \frac{R_S}{Z_S} \left(\frac{Z_S}{Z_{b'e} g_m R_E} + \frac{1}{g_m R_E} + 1 \right) \\ &\simeq \frac{R_S}{Z_S} \left(\frac{Z_S}{Z_{b'e} g_m r_e} + 1 \right) \quad \text{for } g_m R_E > 1 \end{aligned} \tag{10-102}$$

Upon substituting the values of Z_S and $Z_{b'e}$ into Eq. (10-102), the reciprocal of the voltage gain becomes

$$\begin{aligned}\frac{1}{A_v} &\simeq \frac{R_S}{g_m R_E} \frac{1 + s r_{b'e} C_{b'e}}{r_{b'e}} + (1 + s C_{cb'} R_S) \\ &= \frac{R_S}{\beta R_E} (1 + s C_{b'e} r_{b'e}) + (1 + s C_{cb'} R_S) \end{aligned} \tag{10-103}$$

The reciprocal of the voltage gain is made up of the two terms, both of which are a function of frequency. Depending on the relative sizes of the two time constants $C_{b'e} r_{b'e}$ and $C_{cb'} R_S$, it may be possible to neglect one of these terms, thus simplifying the voltage gain to a simple one-break frequency Bode response plot. To illustrate this point, the Bode plots for the two terms in Eq. (10-103) are plotted in Figs. 10-30a and 10-30b, respectively. Notice that the break frequency in the first term corresponds to the beta cutoff. Notice also in Fig. 10-30a that the assumption that $R_S/\beta R_E$ generally is less than 1 is quite valid, since the input impedance into an emitter follower $\simeq \beta R_E$ is usually selected to be larger than the source resistance R_S.

To make use of Figs. 10-30a and 10-30b we need to find ω_1 by first calculating ω_β and $R_S/\beta R_E$. If ω_1 is considerably larger than ω_2 in Fig. 10-30b, the first term is neglected and the upper cutoff frequency ω_H of the emitter follower is approximately ω_2. We can write this symbolically as

$$\omega_H = \frac{1}{R_S C_{cb'}} \quad \text{if } \omega_1 \gg \frac{1}{R_S C_{cb'}} \tag{10-104}$$

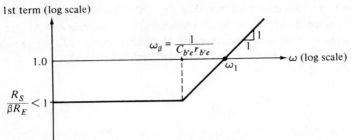

(a) Bodé plot of the first term in the expression for A_v^{-1} in the emitter following amplifier.

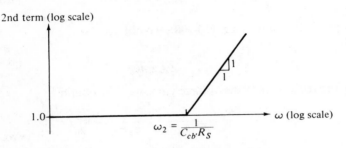

(b) Bodé plot of the second term.

FIGURE 10-30 Bode plots for analysis A_v^{-1} of the CC amplifier.

Equation (10-104) can also be written as follows, if we solve for ω_1,

$$\omega_H = \frac{1}{R_S C_{cb'}} \quad \text{if} \quad \frac{R_E}{R_S}\omega_T \gg \frac{1}{R_S C_{cb'}} \quad (10\text{-}105)$$

A second simplification may arise if ω_2 is much greater than ω_1. In this case, the second term in Eq. (10-103) becomes unity and the expression for voltage gain becomes

$$A_v = \frac{\beta R_E}{(\beta R_E + R_S)\left(1 + sC_{b'e}\dfrac{r_{b'e}R_S}{\beta R_E + R_S}\right)} \quad (10\text{-}106)$$

The upper cutoff frequency ω_H becomes

$$\omega_H = \frac{\beta R_E + R_S}{C_{b'e} R_S r_{b'e}} = \omega_\beta \frac{\beta R_E + R_S}{R_S} \quad \text{if} \quad \frac{1}{C_{cb'} R_S} \gg \frac{R_E}{R_S}\omega_T \quad (10\text{-}107)$$

A third case arises when ω_1 and ω_2 are of the same order of magnitude. In this case, to solve a specific problem we must solve for the voltage gain from Eq. (10-103).

Example 10-4: In the circuit shown in Fig. 10-31 the emitter follower is biased by a previous stage in which the source impedance is essentially resistive and

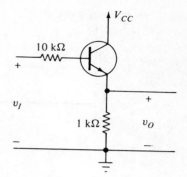

FIGURE 10-31 Emitter follower of example Problem 10-3.

is equal to 10 kΩ. The transistor used has the following parameters associated with it:

$$\beta = 100$$
$$f_T = 400 \text{ MHz}$$
$$C_{cb'} = 3.3 \text{ pF}$$

Estimate the upper cutoff frequency ω_H of the transistor amplifier circuit shown.

Solution:

The β cutoff frequency is

$$f_\beta = \frac{f_T}{\beta} = \frac{400}{100} = 4 \text{ MHz}$$

$$= \frac{1}{C_{b'e} r_{b'e}}$$

To obtain the Bode plot of the first term of Eq. (10-103), we calculate

$$\frac{R_S}{\beta R_E} = \frac{10 \text{ k}\Omega}{(100)(1 \text{ k}\Omega)} = \frac{1}{10}$$

The Bode plot is then as shown in Fig. 10-32a. It is easy to see that $f_1 = 2\pi\omega_1$ is equal to 40 MHz in Fig. 10-32a. The second term of Eq. (10-103) can be calculated by finding $f_2 = \omega_2/2\pi$:

$$f_2 = \frac{1}{2\pi(C_{cb'} R_S)} = \frac{1}{2\pi(3.3 \times 10^{-12} \times 10^4)}$$
$$= 4.8 \text{ MHz}$$

The Bode plot associated with the second term is then as shown in Fig. 10-32b. Since f_2 is much larger than f_1, f_2 is approximately the upper cutoff frequency:

$$f_H = 4.8 \text{ MHz}$$

Generally speaking, the frequency response of the emitter follower is much higher than for a CE amplifier, but, again, this depends upon the source impedance that is used. With careful design it is easy to obtain an upper

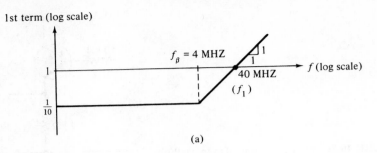
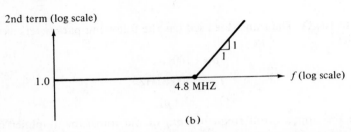

FIGURE 10-32 The Bode plots of two terms of Eq. (10-103) in example problem 10-3: (a) the first term (b) the second term.

cutoff frequency of an emitter follower to operate close to f_T. In this case, our analysis becomes invalid because of the assumption that was made in obtaining Eq. (10-98). If a cascade of a CE amplifier and an emitter follower is used, the gain of the CE amplifier will generally decrease with frequency first, unless the CE amplifier is driven by a very low source impedance.

10.8 Frequency Response of a Field Effect Transistor

The FET at high frequencies can be described by an equivalent circuit in Fig. 10-33b. The capacitances C_{gd} and C_{gs} are a result of the reverse-biased junction in the FET and act in a similar fashion to $C_{cb'}$ in the bipolar transistor.

Typical values of C_{gs} range from 100 pF for a low-frequency transistor to less than 4 pF for a high-frequency transistor. C_{ds}, on the other hand, varies from about 10 pF to less than a fraction of 1.0 pF for high-frequency transistors.

If the transistor is used in the common source configuration as shown in Fig. 10-34a, the equivalent circuit shown in Fig. 10-34b may be used to obtain the upper cutoff frequency f_H of the amplifier. Notice that the Miller effect occurs in the field effect transistor as well. It is easy to see that f_H becomes

$$f_H = \frac{1}{2\pi R_S[C_{gs} + (1 + g_m R'_L)C_{gd}]} \quad \text{where } R'_L = R_L || r_d$$

Sec. 10.8 / Frequency Response of a Field Effect Transistor   279

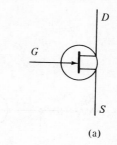

(a)

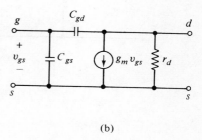

(b)

FIGURE 10-33 The FET and high-frequency equivalent circuit.
(a) An n channel FET.
(b) A typical FET high-frequency equivalent circuit.

Typical values for f_H run much lower for an FET in the common source configuration than for a CE amplifier using a transistor of similar cost, especially if the FET is used with a very high source impedance.

The frequency response of the source follower and the common gate amplifier is left as an exercise for the student.

Example 10-5: The common source amplifier shown in Fig. 10-34 has the following circuit component values:

$$R_L = 5 \text{ k}\Omega$$
$$R_S = 100 \text{ k}\Omega$$
$$R_g = 2 \text{ M}\Omega$$

The transistor in the circuit has the following parameters:

$$r_d = 100 \text{ k}\Omega$$
$$C_{gs} = 50 \text{ pF}$$
$$C_{gd} = 5 \text{ pF}$$
$$g_m = 5 \times 10^{-3} \text{ mho}$$

Obtain f_H for this amplifier.

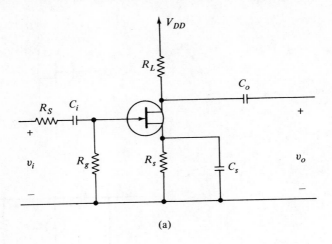

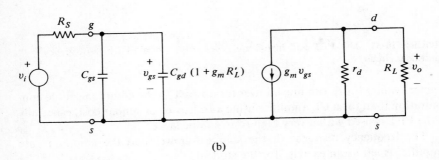

(a) A FET amplifier
(b) An equivalent circuit for the FET amplifier using the high frequency FET model and the Miller effect

FIGURE 10-34 FET common source amplifier and equivalent circuit.

Solution:

$$f_H \simeq \frac{1}{2\pi R_S[C_{gs} + (1 + g_m R'_L)C_{gd}]}$$

Now

$$g_m R'_L \simeq g_m R_L = 25$$

so that

$$f_H = \frac{1}{2\pi \times 10^5(1.80 \times 10^{-10})}$$
$$= 88 \text{ kHz}$$

PROBLEMS

10.1 If $h_{ie} = 5$ kΩ, $h_{fe} = 100$, $1/h_{oe} = 50$ kΩ and $h_{re} = 4 \times 10^{-4}$, find $r_{b'e}$, r_{ce}, $r_{cb'}$ and g_m. Assume that $r_{bb'} = 20$ Ω.

10.2 Assume the effect of h_{re} to be negligible and find $r_{b'e}$, r_{ce}, and g_m for the values given in Problem 10.1.

10.3 Given the following r-parameters for a transistor, $r_e = 26$ Ω, $r_b = 20$ Ω, and $r_d = 10$ kΩ, find the low-frequency hybrid-pi parameters.

10.4 A transistor manufacturer specifies $C_{ob} = 2.0$ pF and $f_T = 700$ MHz. What additional information would be needed in order to determine $C_{cb'}$ and $C_{b'e}$?

10.5 (a) If the β-cutoff of a transistor is known to be 1 MHz and $β_o = 100$, sketch the frequency response of β as a function of frequency. What will f_T be for this transistor?
(b) What would be the α-cutoff for this transistor?

10.6 Make a sketch of a graph showing the relationships between $C_{b'e}$ and I_E when I_E varies from 1 mA to 4 mA. Assume that $m = 1.4$ and $f_T = 50$ MHz.

10.7 The silicon transistor in Fig. P10-7 has the following parameters given by the manufacturer: $C_{ob} = 4.0$ pF, $f_T = 100$ MHz, $h_{fe} = 60$, and $r_{bb'} = 20$ Ω. Obtain the upper frequency cutoff f_H for the gain of this amplifier.

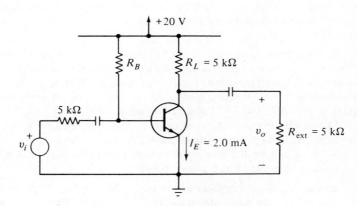

FIGURE P10-7

10.8 (a) In Fig. P10-7, the resistor R_{ext} is changed to 1.0 kΩ. What will be the new upper cutoff frequency f_H' of the amplifier?
(b) Show how an experiment could be set up in the laboratory to determine $C_{cb'}$ and $C_{b'e}$ by changing R_{ext} in Fig. P10-7, as was done in part (a) of this problem. Comment on the accuracy of such a method.

10.9 (a) Sketch the Bode plots for the voltage gains v_o/v_i of Problems 10.7 and 10.8 on the same graph.
(b) Find the frequency at which the gain is approximately equal to unity in each case in part (a).

10.10 Show how the Miller effect can be modified for use in the mid-band frequency region for the circuit shown in Fig. P10-10. Assume $C_{cb'}$ and $C_{b'e}$ to be small and C_i and C_o to be large. Use this method to obtain formulas for the input impedance R_{in} and the voltage gain v_o/v_i.

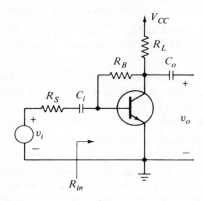

FIGURE P10-10

10.11 Bias the amplifier in Fig. P10-10 and calculate the input R_{in} and voltage gain for your circuit. How do these results compare with that of a constant current biased CE amplifier using the same transistor and the same R_S and R_L?

10.12 A CB amplifier is capacitively coupled to both a source and load impedance. The source impedance is 50 Ω and the load impedance 500 Ω. A designer builds a CB amplifier with an input impedance of 50 Ω at 1 kHz and has a voltage gain of 5.0 when connected to these loads. A transistor with an f_T of 700 MHz and a C_{ob} of 1.5 pF is used.
(a) Sketch a Bode plot of the resulting voltage gain as function of frequency.
(b) What is the upper cutoff frequency of this amplifier?
(c) What would be the new frequency response if a transistor with the same f_T were used but C_{ob} were 2.5 pF?

10.13 (a) Design a cascode amplifier of the type shown in Fig. 10-27. Use transistors with an f_T of 400 MHz and $C_{ob} = 4$ pF. The values R_S and R_L are specified as 2 kΩ and 4.7 kΩ, respectively. Determine the upper frequency cutoff of the amplifier.
(b) How does the frequency response in part (a) compare if a CE amplifier were used with the same R_S and R_L and biased at the same Q-point as T_2 in Fig. 10.27?

10.14 Determine the voltage gain of the EF shown in Fig. 10-31, using the same transistor as in Example 10-3 at 30 MHz.

10.15 (a) If the dc emitter current I_E in the emitter follower was increased by

30 per cent in Fig. 10-31, how would this affect the frequency response of the amplifier?

(b) If only the dc emitter current in a CE amplifier were changed, would this affect the frequency response? Why?

10.16 Determine the upper cutoff frequency of the amplifier shown in Fig. P10-16.

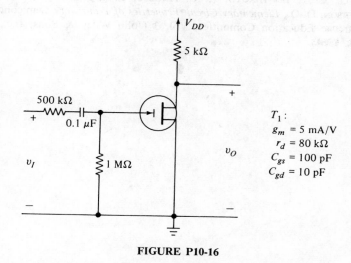

FIGURE P10-16

10.17 Derive the formula for the high-frequency response of a common gate amplifier.

10.18 Determine the upper frequency cutoff of the amplifier shown in Fig. P10-18. Assume r_d to be large.

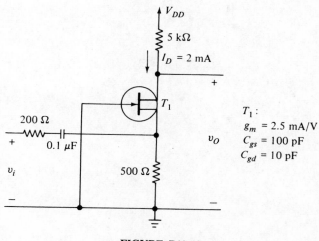

FIGURE P10-18

REFERENCES

[1] GIBBONS, J. F., *Semiconductor Electronics* (McGraw-Hill Book Co., New York, 1966).

[2] SEARLE, C. L., BOOTHROYD, A. R., ANGELO, E. J., JR., GRAY, P. E., and PEDERSON, D. O., *Elementary Circuit Properties of Transistors*, Semiconductor Electronic Education Committee, Vol. 3 (John Wiley & Sons, Inc., New York, 1964).

11

POWER SUPPLIES

In all the amplifiers designed so far in this text, there has been an abundant use of the battery symbol. This symbol may represent a physical battery but more often represents a constant (or dc) voltage source provided by a power supply. Since transistors are used in the circuit design of dc power supplies and since dc power supplies are so necessary for every circuit, we shall devote this chapter to the design of simple power supplies.

The following is not intended as a complete study of power supplies, since a whole text could be devoted to this topic. We shall introduce some simple power supply circuits that can be effectively used to supply a dc source for the many amplifier circuits studied in this text.

The main object in designing power supplies for transistor circuits is to obtain as constant a dc supply as possible from the readily available household sources, which are usually 115-V rms sinusoidal ac sources, operating at a frequency of 60 Hz.

11.1 Rectification

The idea of rectification has been already introduced in Chapter 5 as one of the applications of a semiconductor diode. The circuit of Chapter 5 is repeated in Fig. 11-1a with the addition of a transformer to change the value of the ac source to a more desirable voltage level. This circuit is called a *half-wave rectifier*. The voltage v_i represents the secondary voltage of the transformer and is given by

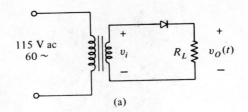

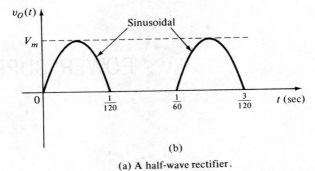

(a) A half-wave rectifier.
(b) The output voltage of an ideal half-wave rectifier.

FIGURE 11-1 Circuit and wave form of a half-wave rectifier.

$$v_i = V_m \sin \omega t \qquad (11\text{-}1)$$

where $\omega \simeq 377$ rad/s

If the diode is assumed to be ideal, then the output voltage obtained is that shown in Fig. 11-1b.

The average value of $v_o(t)$, V_{AVG}, can now be calculated as follows:

$$V_{AVG} = \frac{\int_0^{1/120} V_m \sin 377 t \, dt}{\frac{1}{60}} \qquad (11\text{-}2)$$

$$= \frac{V_m}{\pi}$$

The average voltage of $v_o(t)$ is the dc component of the output voltage, since v_o can be thought of as being made up of two parts: the average value and the ac component of $v_o(t)$. In this case the ac component has a peak-to-peak value of V_m. The ac component of $v_o(t)$ is shown in Fig. 11-2. Thus the output voltage v_o can be written as the sum of the dc or average value, plus the ac component $v_o(t)_{ac}$:

$$v_o(t) = V_{AVG} + v_o(t)_{ac} \qquad (11\text{-}3)$$

In building power supplies we shall seek ways in which to reduce $v_o(t)_{ac}$ to zero while maintaining V_{AVG}. In a rough manner of speaking, V_{AVG} repre-

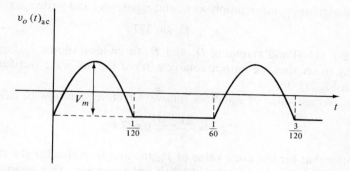

FIGURE 11-2 The ac component of a half-wave rectified sinusoidal voltage.

sents the useful part of $v_o(t)$. Notice the relative size of V_{AVG}, as given by Eq. (11-2).

The relative size of V_{AVG} can be increased by the use of a center-tapped transformer to construct a *full-wave rectifier*, as shown in Fig. 11-3a. Since a center-tapped transformer is used, it is easy to see that

$$v_{i1} = -v_{i2} \qquad (11\text{-}4)$$

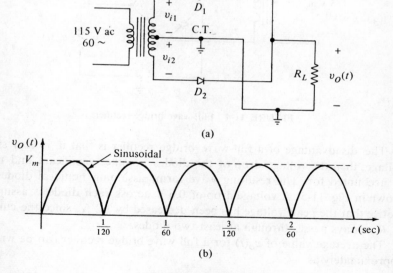

(a) A full-wave rectifier.
(b) The output voltage of a full-wave rectifier.

FIGURE 11-3 Circuit and output voltage of a full-wave rectifier.

Let us designate v_{i1} more simply as v_i, and v_{i2} as $-v_i$, and write v_i as

$$v_i = V_m \sin 377 t \tag{11-5}$$

Using Eq. (11-4) and assuming D_1 and D_2 to be ideal diodes in Fig. 11-3a, it is easy to see that the output voltage $v_o(t)$ of the full-wave rectifier is that given in Fig. 11-3b.

The average value of $v_o(t)$ in a full-wave rectifier can now be calculated:

$$V_{AVG} = \frac{2V_m}{\pi} \simeq 0.637 V_m \tag{11-6}$$

Notice that for the same value of V_m the average value for the full-wave rectifier is twice as large as for the half-wave rectifier. This means that we have a larger dc voltage component at the output. The ac component still has a peak-to-peak value of V_m.

There is another way to construct a full-wave rectifier, without the use of a center-tapped transformer, and that is to use a *full-wave bridge rectifier*, as shown in Fig. 11-4. If $v_i(t)$ is the same as given by Eq. (11-5) and ideal diodes are assumed, then the resulting wave form for $v_o(t)$ is the same as that given in Fig. 11-3b.

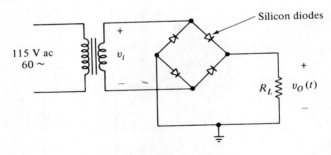

FIGURE 11-4 Full-wave bridge rectifier.

The disadvantage of a full-wave bridge rectifier is that if v_i is a small voltage, then the voltage drop of the diodes becomes significant and must be accounted for. The resulting wave form, assuming nonideal diodes, is shown in Fig. 11-5. A voltage drop of 0.6 V across each diode is assumed. Notice that the peak voltage has been decreased by 1.2 V, since the current to R_L always passes through at least two diodes.

The average value of $v_o(t)$ for a full-wave bridge rectifier can be written approximately as

$$V_{AVG} \simeq \frac{2}{\pi}(V_m - 1.2) \text{ V} \tag{11-7}$$

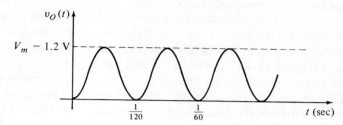

FIGURE 11-5 The output voltage of a full-wave bridge when silicon diodes are used.

11.2 Ripple and Regulation

In order to be able to measure the quality of the dc output voltage of a rectifier, or power supply, two terms—*ripple* and *regulation*—are defined. Per cent ripple is usually defined as

$$\text{per cent ripple} = \frac{\text{rms value of ac component of } v_O(t)}{\text{average value of the output voltage}} \times 100 \text{ per cent} \quad (11\text{-}8)$$

This definition is a standard definition but will not be used very often, since the rms value of the ac component is difficult both to calculate and to measure. Instead we shall define a ripple factor r as follows:

$$r = \frac{\text{peak-to-peak component of the output voltage}}{\text{average value of the output voltage}} \quad (11\text{-}9)$$

Equation (11-9) has the advantage that the peak-to-peak output voltage can be quickly estimated and easily measured by an oscilloscope in the laboratory. The average value, on the other hand, can be measured by a dc or average responding voltmeter.

If the ripple factor r is calculated for the half-wave rectifier, we see that it is much larger than unity and about 1.6 for the full-wave rectifier. For a practical dc power supply, this value should be reduced to much less than 0.1.

The regulation of a power supply, or rectifier, on the other hand, is defined as

$$\text{per cent regulation} = \frac{V_{AVG}(\text{no load}) - V_{AVG}(\text{full load})}{V_{AVG}(\text{full load})} \times 100 \text{ per cent}$$

$$(11\text{-}10)$$

where $V_{AVG}(\text{no load})$ is the average value of the output voltage when the external load resistance is infinite, and $V_{AVG}(\text{full load})$ is the average value of the output voltage when the external load resistance is at a minimum.

The per cent regulation is an indication of how the output voltage changes because of load. In a practical rectifier circuit, such as in a full-wave rectifier, the average output voltage will decrease as the load increases, since there will be a voltage drop in the transformer-winding resistance and an additional voltage drop in the diodes as current through the diodes increases.

A well-designed power supply or rectifier circuit should have a low-regulation factor. This can be achieved mainly by using transformer with low-winding resistance. There is very little that can be done about the increased voltage drop in the diode except to use voltage regulators. These will be discussed in a later section.

11.3 The R–C Filtered Rectifier

To decrease the ripple factor r of a rectifier, an R–C filter is used, as shown in Fig. 11-6a. This results in an output voltage $v_o(t)$, as shown in Fig. 11-6b. This becomes our first circuit for a practical power supply.

The operation of the circuit is relatively straightforward. During the time between t_0 and t_1 (t_0 is the time when the power supply is turned on), the

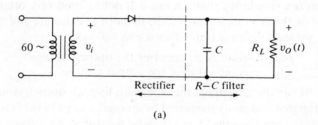

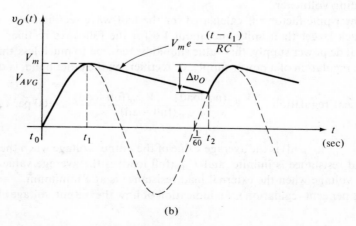

FIGURE 11-6 Half-wave power supply and output voltage.

diode is forward biased and acts as a short circuit. The output voltage $v_o(t)$ is the same as v_i and reaches a maximum value of V_m at t_1.

The voltage across the capacitor C has also reached this value, since v_i is applied directly across it. After the time t_1, v_i begins to decrease below the capacitor voltage, causing the diode to become reverse biased. The output voltage is now determined by the parallel combination of R_L and C only, since the diode has become an open circuit. During the time between t_2 and t_1, C discharges into R_L with a time constant $R_L C$. The value of the voltage in this time period is given by

$$v_o(t)\bigg|_{t_1 < t < t_2} = V_m e^{-(t-t_1)/RC} \tag{11-11}$$

At time t_2 the diode becomes forward biased and the process is repeated.

The ripple factor is given by

$$r = \frac{\Delta v_o}{V_{AVG}} \tag{11-12}$$

where Δv_o is the peak-to-peak output voltage of Fig. 11-6b and V_{AVG} is the average value of $v_o(t)$ of Fig. 11-6b. The ripple factor as given by Eq. (11-12) can now be calculated by making a series of approximations. Since r is usually quite small for practical power supplies, V_{AVG} is approximated by V_m. Example 11-1 illustrates these approximations more clearly.

Example 11-1: A power supply consisting of a half-wave rectifier and R–C filter is constructed as shown in Fig. 11-7. The voltage v_i is given by

$$v_i = 100 \sin 377t$$

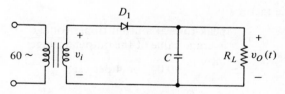

FIGURE 11-7 Circuit of Ex. 11-1.

and D_1 is a silicon diode. The component values are

$$C = 400 \ \mu F$$
$$R_L = 1{,}000 \ \Omega$$

Calculate the peak-to-peak ripple voltage at the output and the ripple factor r, assuming the transformer in Fig. 11-7 has no winding resistances.

Solution:

The resulting output voltage is sketched in Fig. 11-8. The voltage in the time interval between $t = 0$ and t_1 is given by

$$v_o(t) = 100 e^{-(1/RC)t} \tag{11-13}$$

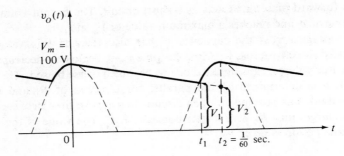

FIGURE 11-8 Output wave form of the filtered rectifier in Ex. 11-1.

To calculate the peak-to-peak ripple voltage, we need to find the value of V_1. To do this we must first find the value for t_1 and substitute this into Eq. (11-13) to solve for the voltage at t_1. However, this is quite a lengthy exercise in algebra, so we shall assume V_1 to be approximately equal to V_2 and calculate the value of V_2 by substituting the value of t_2 into Eq. (11-13). V_2 is actually smaller than V_1 and the calculated ripple will be larger than the actual ripple; the error, however, will not be that large.

$$V_1 \simeq V_2 = 100 e^{-(1/RC)t_1}$$
$$= 100 e^{-(1/10^3 \times 400 \times 10^{-6})(1/60)}$$
$$= 100 e^{-0.042} \simeq 96 \text{ V}$$

The peak-to-peak ripple voltage Δv_o becomes

$$\Delta v_O \simeq V_m - V_2 = 4 \text{ V peak to peak}$$

and the ripple factor r is

$$r = \frac{\text{peak-to-peak output ripple voltage}}{\text{average value of the output voltage}}$$
$$\simeq \frac{4 \text{ V}}{100 \text{ V}} = 0.04, \text{ or 4 per cent}$$

From the above example it is apparent that to decrease the ripple factor r one only need increase C, thus making the ripple as small as one would want. However, this practice may lead to some very large sizes of capacitors needed for the power supply. Hence, the designer is usually limited by the physical size of the capacitor. The ripple in Ex. (11-1) can also be decreased by using a full-wave rectifier rather than a half-wave rectifier. However, probably the greatest disadvantage is that the ripple and regulation change with the load resistance. To overcome this the designer should design for the maximum allowable ripple to occur at the lowest value of load resistance. Any decrease in the load would decrease the ripple.

The power supply obtained by using an R–C filtered rectifier is a good simple power supply and can be used for many applications, especially where high currents are not needed. They are relatively inexpensive, require a minimum of parts, and are used in many commercial instruments, such as transistor radios and low-powered audio amplifiers.

11.4 Emitter Follower Voltage Regulator

To decrease the ripple and improve the regulation in R–C filtered rectifiers, voltage regulators are used. The simplest of these is the emitter follower voltage regulator circuit, shown in Fig. 11-9. The operation of this type of

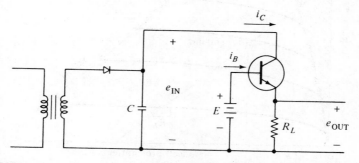

FIGURE 11-9 A half-wave rectifier with an emitter follower voltage regulator.

regulator is relatively simple. The output voltage e_{OUT} is given by

$$e_{OUT} = E - v_{BE}$$
$$\simeq E - 0.7 \qquad (11\text{-}14)$$

Notice that the output voltage e_{OUT} is simply the battery voltage E minus the base-to-emitter voltage v_{BE}, which remains approximately constant as R_L is changed. Naturally, the voltage e_{IN} should always be larger than E to keep transistors properly biased. The current required from the battery is i_B and is given by

$$i_B = \frac{i_C}{h_{FE}} \qquad (11\text{-}15)$$

Since i_C is approximately the current to the load, notice that the load on the battery has been decreased by a current gain of the transistor h_{FE}. To further decrease this drain a two-stage Darlington connected emitter follower may be used.

The presence of the battery in the circuit may be undesirable, and the need for one will be eliminated later in this chapter.

A load line analysis of the emitter follower regulator gives a better insight into its operation. Consider the load line shown in Fig. 11-10. Notice that the intersection of the load line of the v_{CE} axis is the voltage across the filter capacitor C, and it is itself varying because of ripple. This means that

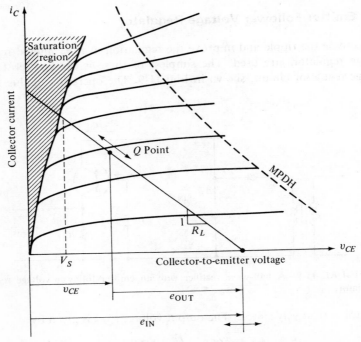

FIGURE 11-10 Load line description of the operation of an emitter follower voltage regulator.

the load line moves back and forth with a constant slope of $-1/R_L$. Since e_{OUT} is approximately constant, it is important that v_{CE} remain out of the saturation during the entire operation for the transistor to operate correctly. This can be done by ensuring that v_{CE} is larger than the saturation V_S during all values of e_{IN} and R_L. v_{CE} will be in the linear operating region of the transistor characteristics at all times. A final point in the design of an emitter follower is to ensure that the Q-point is within the safe operating region of the transistor.

The next question that arises is: How much of the ripple voltage that appears across C appears at the output? In Fig. 11-10, e_{OUT} was assumed to be constant and all the variations were assumed to be taken up by the voltage v_{CE}. This is not entirely true and a small-signal analysis is necessary to determine the improvement in the ripple factor at the output. Fig. 11-11a shows the essential portion of the power supply circuit of Fig. 11-9 to answer this question.

Figure 11-11b is the small-signal equivalent circuit of Fig. 11-11a, where e_i is the variational portion of e_{IN}, which is the ripple voltage across the

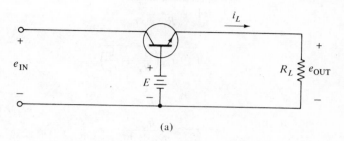

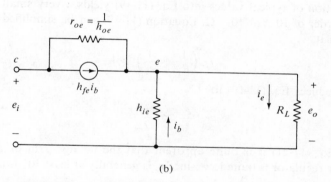

FIGURE 11-11 Regulator circuit and equivalent circuit.

capacitor, and e_o represents the ripple that appears at the output. The parameter h_{re} has been neglected. For this circuit to be effective the voltage gain or the ratio of e_o to e_i should be as small as possible.

Using Kirchhoff's current law at node e, we can write

$$\frac{e_o}{R_L} = (e_i - e_o)h_{oe} + (h_{fe} + 1)i_b \qquad (11\text{-}16)$$

The base current i_b is given by

$$i_b = \frac{-e_o}{h_{ie}} \qquad (11\text{-}17)$$

Substituting Eq. (11-17) into Eq. (11-16), we obtain

$$\frac{e_o}{R_L} = (e_i - e_o)h_{oe} - (h_{fe} + 1)\frac{e_o}{h_{ie}} \qquad (11\text{-}18)$$

From Eq. (11-18), we obtain the expression for the voltage gain to be

$$\frac{e_o}{e_i} = \frac{h_{oe}}{\frac{1}{R_L} + \frac{h_{fe}+1}{h_{ie}} + h_{oe}}$$

$$= \frac{1}{1 + \frac{r_{oe}}{R_L} + \frac{r_{oe}}{r_e}} \qquad (11\text{-}19)$$

$$= \frac{1}{1 + r_{oe}\left(\frac{1}{R_L} + \frac{1}{r_e}\right)}$$

where
$$r_{oe} = \frac{1}{h_{oe}}$$

and
$$r_e \simeq \frac{h_{ie}}{h_{fe}+1}$$

Substitution of typical values into Eq. (11-19) yields a very small quantity, of the order of 10^{-3} or 10^{-4} Ω. Equation (11-19) can be simplified by recognizing that

$$r_{oe}\left(\frac{1}{R_L} + \frac{1}{r_e}\right) \gg 1 \qquad (11\text{-}20)$$

which reduces Eq. (11-19) to

$$\frac{e_o}{e_i} \simeq \frac{R_L \| r_e}{r_{oe}} \qquad (11\text{-}21)$$

From Eq. (11-21) it is quite apparent that the voltage gain of an emitter follower regulator is quite low, since r_e is generally at least 10^3 times smaller than r_{oe}.

11.5 Zener Diode Regulator

A *zener diode* is a *pn* diode that has the characteristics shown in Fig. 11-12a. The symbol and the current and voltage reference directions are shown in Fig. 11-12b. In the forward direction ($v_D > 0$) the zener diode acts the same as a conventional diode, as well as in the reverse direction for voltages greater than $-E_Z$. At $-v_D = E_Z$, the *zener voltage*, the diode breaks down without destroying itself and acts almost as a short circuit with a very low dynamic resistance. The dynamic resistance r_d in this region is shown in Fig. 11-12a and is the reciprocal of the slope of the v–i characteristic. Typical values of r_d range from a few ohms to about 30 Ω with a typical value of about 20 Ω.

It is in this breakdown region that the zener diode is used. The amount of current that can be passed through the diode in the reverse direction is

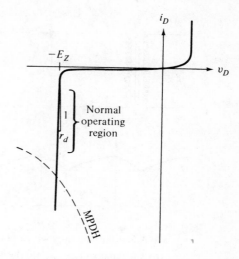

FIGURE 11-12 Zener diode; characteristics and diode symbol. (a) Zener diode $v-i$ characteristics. (b) Symbol, voltage, and current reference directions.

limited by the power rating of the diode, which simply is the product of $-E_Z$ and the maximum reverse current.

If the power rating is given as P_d, then a maximum dissipation hyperbola can be drawn, as shown in Fig. 11-12a, and the diode can be operated as long as the following expression is satisfied:

$$P_d > v_D i_D \qquad (11\text{-}22)$$

Zener diodes can be obtained for various zener voltages ranging from a fraction of 1 V to about 100 V. The power rating also varies from about 1/4-W to as high as 100 W.

Zener diodes can be used to make effective voltage regulator circuits. Fig. 11-13a illustrates a power supply using a zener diode regulator circuit. The zener diode regulator consists of R, R_L, and zener diode D_2. D_1 is a conventional silicon diode.

The operation of this regulator is straightforward. If e_{IN} is the voltage, as given by Fig. 11-13b, and E_{MAX} is the maximum and E_{MIN} the minimum value of e_{IN}, then R is chosen so that there is enough current flow through

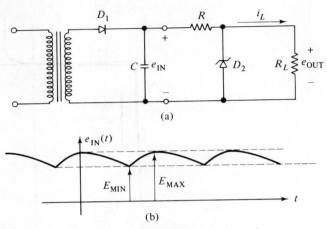

(a) A half-wave rectifier with a zener diode regulator.
(b) The input voltage e_{IN} into the zener diode regulator

FIGURE 11-13 Half-wave regulator and its input voltage.

to reverse-bias the diode D_2 at $-E_z$ during all time of e_{IN}, including when e_{IN} is at a minimum value E_{MIN}.

To calculate R in Fig. 11-13a, a minimum value of R_L as well as the voltage e_{IN} must be known. When R_L is at a minimum, the maximum load current i_L flows, while just enough current flows through D_2 to keep it properly biased in the zener operating region of the diode. This current through the zener diode will drop to a minimum when e_{IN} is at a minimum. It is under this condition that the R is calculated. When the voltage e_{IN} goes up, then more current flows through the zener diode. The required power rating P_d of the diode is calculated when the maximum current flows through D_2. This occurs when R_L is infinite and e_{IN} is at a maximum value.

The procedure for designing a zener diode regulator is illustrated by means of the following example.

Example 11-2: A voltage e_{IN} is available that varies in the manner shown in Fig. 11-14a. A power supply using the zener diode regulator shown in Fig. 11-14b is to be designed that is capable of delivering a dc load current i_L up to 10 mA, at a voltage of 10 V.

Solution:

It is interesting to know the *minimum* value of R_L

$$R_L = \frac{\text{dc output voltage}}{\text{maximum output current}}$$

$$= \frac{10 \text{ V}}{10 \text{ mA}} = 1.0 \text{ k}\Omega$$

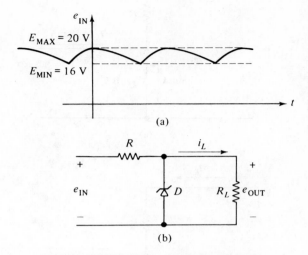

(a) Input wave-form of a zener diode regulator for Ex. 11-2
(b) The zener diode regulator for Ex. 11-2

FIGURE 11-14 Waveform and circuit of Ex. 11-2.

Thus the load R_L can be described as being

$$1.0 \text{ k}\Omega < R_L < \infty$$

The next step is to choose a 10-V zener diode for the circuit. (The power rating is to be determined later.)

To determine R we now consider the case where current through the zener diode D is at a minimum (see Fig. 11-15a). This occurs when e_{IN} is at a minimum, in this case, 16 V, and i_L is at a maximum, at 10 mA. We choose the minimum current through D to be -5 mA to ensure that the zener diode is biased well into the linear region, as shown in Fig. 11-15c. This condition is shown as P_1. The voltage across R is 6 V and the current through R is 15 mA. Thus R should be

$$R = \frac{6 \text{V}}{15 \text{ mA}} = 400 \Omega$$

If i_L were suddenly removed, then the current through D would rise to 15 mA. This is shown as P_2 in Fig. 11-15b.

To determine the power rating necessary for the zener diode we now calculate the maximum current through D. This occurs when e_{IN} is at a maximum and i_L is at a minimum, as illustrated in Fig. 11-15c. At this point the current through D is equal to the current to R. The voltage across R is 10 V. Thus the current through D can be calculated:

$$-i_D = i_R = \frac{10 \text{ V}}{400 \text{ }\Omega} = 25 \text{ mA}$$

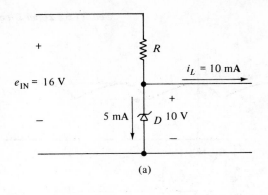

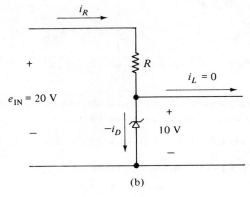

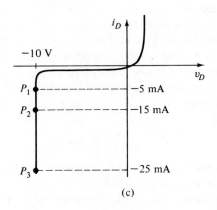

(a) Operating condition of regulator during maximum load and minimum input voltage.
(b) Operation during minimum load and maximum input voltage.
(c) Operating points of $v - i$ curve of zener diode.

FIGURE 11-15 Solution to Ex. 11-2.

This condition is shown in Fig. 11-15b as point P_3. To ensure that the diode has a sufficient power rating, we calculate the power dissipation at P_3 and arrange for the maximum power dissipation hyperbola to lie away from this point:

$$P_d = i_D v_D \text{ at } P_3$$
$$= (-25 \text{ mA})(-10 \text{ V}) = 0.25 \text{ W}$$

To ensure long life in the circuit a 1/2-W zener diode should be used in this circuit, as opposed to a 1/4-W diode (which would work quite well, since average power dissipation is probably lower than the power at point P_3 in Fig. 11-15b). The cost difference would probably be minimal anyway.

Theoretically it is possible to make the diode current at P_1 as low as possible. However, the ratio of P_3 to P_1 should be not too large, since variations in the resistance R due to heat and tolerances may be enough to reduce voltage across the zener diode below the zener voltage, if the minimum current through the zener diode is made too small.

In order to determine the ripple at the output of the zener diode regulator, a small-signal analysis should be performed. The regulator circuit with its small-signal circuit is shown in Figs. 11-16a and 11-16b. The dynamic resistance r_d of D is defined as in Fig. 11-12. The ratio of the output small-signal voltage and the input small-signal voltage can be calculated from Fig. 11-16b

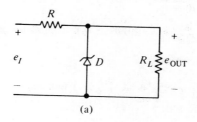

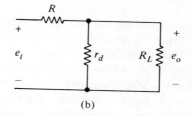

(a) A zener diode regulator circuit.

(b) The small signal equivalent circuit of the regulator.

FIGURE 11-16 Zener diode regulator and equivalent circuit.

by voltage division to be

$$\frac{e_o}{e_i} = \frac{R_L || r_d}{R + (r_d || R_L)} \qquad (11\text{-}23)$$

If r_d is much less than R_L, then Eq. (11-23) becomes

$$\frac{e_o}{e_i} \simeq \frac{r_d}{R + r_d} \qquad (11\text{-}24)$$

It is interesting to note that the voltage ratio is reduced in Eq. (11-24) by having R as large as possible. Since R is determined by minimum current through the zener diode, it is wise to make this as small as possible so that R is as large as possible. Notice also that R becomes larger if the value of e_I is much larger than the value of e_{OUT}. This means that the ripple at the output is small when a small dc output voltage e_{OUT} is derived from a large input voltage e_I.

Example 11-3: Given the circuit of Example 11-2 and the dynamic resistance of D as $10\,\Omega$, calculate the output ripple voltage and the ripple factor r.

Solution:

$$e_i = 4 \text{ V peak to peak} \qquad \text{(from Fig. 11-14a)}$$

Thus the output ripple voltage becomes

$$e_o \simeq \frac{r_d}{R + r_d} e_i$$

$$= \frac{10}{410}(4) = 0.098 \simeq 0.1 \text{ V peak to peak}$$

The ripple factor r becomes

$$r \simeq \frac{e_o}{E_{OUT}} = \frac{0.1}{10} \simeq 1 \text{ per cent}$$

11.6 Regulated Power Supply Design

In Section 11-4 we have already seen a very good regulated power supply. The emitter follower regulator provided a relatively ripple-free output voltage. The greatest disadvantage of the circuit is the need for a battery. In this section we shall try to show how this battery can be eliminated.

From Fig. 11-9 in Section 11-4 it is apparent that a regulated power supply consists of several major stages or sections. These sections are the rectifier, the filter, voltage reference, control element, and the load. A block diagram of a typical power supply circuit is shown in Fig. 11-17. Various combinations of the different types of sections can be used. For example, for the rectifier circuit we may use a full-wave, half-wave, or bridge rectifier. For the filter, we have only studied the capacitive type. For a reference voltage we may

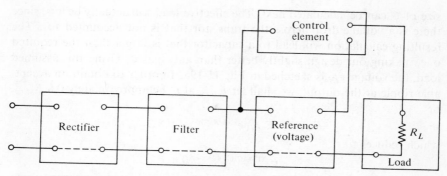

FIGURE 11-17 Block diagram of a voltage regulator using a reference.

use a battery or probably a zener regulated supply. In the control element we can use an emitter follower. There are other possibilities, of course, that we have not studied. One of the possible combinations of the circuits we have studied is shown in Fig. 11-18. This circuit employs a half-wave rectifier, capacitive filter, zener diode reference voltage, and an emitter follower as the control element. The load is represented by R_L.

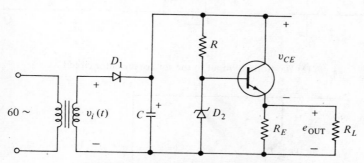

FIGURE 11-18 A regulated power supply.

Example 11-4: The circuit of Fig. 11-18 has been selected as a possible choice for use as a power supply.

(a) Design a power supply using the given circuit, capable of delivering 0.5 A at 12 V.

(b) Estimate the ripple of the resulting power supply.

Solution:

(a) Calculate the minimum value of R_L:

$$R_L = \frac{12 \text{ V}}{0.5 \text{ A}} = 24 \text{ }\Omega$$

We shall assume R_L to be the effective load across the capacitor C so that the size of C can be calculated next. The effective load will actually be less, since there is a voltage drop across the transistor that is not accounted for. The resulting calculation will lead to a capacitor that is larger than the required one, making our design slightly better than anticipated. Using this assumed load, the voltage e_{IN} is sketched in Fig. 11-19a. In order to obtain an acceptable ripple at the output, we shall let E_{MIN} at t_1 be approximately $0.8\,E_{MAX}$, or

$$E_{MIN} = E_{MAX} e^{-(1/R_L C)t_1} = 0.8 E_{MAX}$$

which reduces to

$$e^{-(1/24C)(1/60)} \simeq 0.8$$

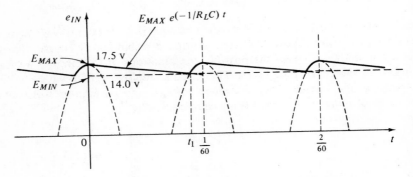

(a) Waveform of input to the voltage regulator circuit

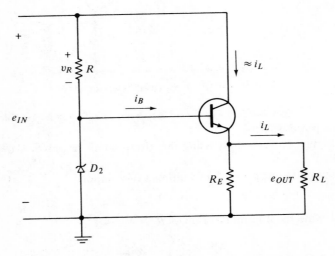

(b) The voltage regulator portion for the circuit used in Ex. 11-4.

FIGURE 11-19 Figures illustrating solution of Ex. 11-4.

Solving for C, we obtain

$$C \simeq 3{,}200 \ \mu\text{F}$$

When e_{IN} has reached a minimum value, we shall let v_{CE} at this instant be 2 V so that the transistor is adequately biased in the linear region. This makes the voltage E_{MIN} equal to 14 V.

$$E_{MIN} = 14 \text{ V}$$

Now we can solve for E_{MAX}, since $E_{MIN} = 0.8 E_{MAX}$, so that E_{MAX} becomes

$$E_{MAX} = \frac{E_{MIN}}{0.8} = 17.5 \text{ V}$$

If we allow a 0.6-V drop across the diode D_1, we can now calculate $v_i(t)$:

$$\begin{aligned} v_i(t) &= (E_{MAX} + 0.6) \sin 377t \\ &= 18.1 \sin 377t \text{ V} \end{aligned}$$

or

$$V_i = \frac{18.1}{\sqrt{2}} = 12.8 \text{ V rms}$$

This gives us the secondary voltage of the transformer. Probably a 12.6-V filament transformer with low winding resistance can be used.

Under maximum load conditions i_L will be 0.5 A. The maximum base current $(i_B)_{MAX}$ will be 5 mA if a transistor with $h_{FE} = 100$ is used. That is,

$$(i_B)_{MAX} = \frac{(i_L)_{MAX}}{h_{FE}} = \frac{500 \text{ mA}}{100} = 5 \text{ mA}$$

This base current will be supplied by the zener regulator, as shown in Fig. 11-19b.

To calculate R we should choose a 12.6-V zener diode so that output voltage will be approximately 12 V. If under maximum load conditions 5 mA is allowed to flow through the zener diode, then R can be calculated as was done in Example 11-2:

$$R = \frac{E_{MIN} - E_Z}{10 \text{ mA}} = \frac{(14.0 - 12.6) \text{ V}}{10 \text{ mA}} = 140 \ \Omega$$

The maximum current through D_2 will occur when e_{IN} is at its maximum value and i_B is approximately zero. The voltage across R, v_R, at this instant is

$$v_R = 17.5 - 12.6 = 4.9 \text{ V}$$

The current through R, i_R, becomes

$$i_R = \frac{4.9}{140} = 35 \text{ mA}$$

which is the maximum zener diode current. The maximum power dissipation P_d of the zener diode is

$$P_d = -i_R E_Z = 35 \times 10^{-3} \times 12.6 = 440 \text{ mW}$$

A 1/2- or 1-W zener diode should be used.

The instantaneous power in the transistor p_C is given by

$$p_C = v_{CE} i_C$$

Since the maximum value of v_{CE} is 5.5 V, occurring when e_{IN} is at a maximum and the maximum collector current is 0.5 A, it would be safe to choose a transistor with a power rating in excess of the maximum instantaneous power. This again will result in an overestimated power rating but will provide a good safety factor for the transistor by keeping the operating point of the transistor always with the maximum power dissipation hyperbola.

The maximum instantaneous power becomes

$$(p_C)_{max} = 5.5 \times 0.5 = 2.75 \text{ W}$$

A power transistor with a collector power rating in excess of this value should be chosen.

Since a maximum 6.0 W of power is dissipated, and about 3.0 W in the transistor, the transformer chosen for the circuit should be capable of providing about 0.75 A. Finally, a choice for R_E of about 1 or 2 kΩ completes the design of the power supply. This resistor helps to keep the output voltage at approximately 12 V, should R_L be disconnected or reduced to a very low value.

(b) To estimate the ripple in the output voltage, we recognize that almost all the ripple voltage at the output is due to ripple across the zener diode D_2. Since the voltage gain of an emitter follower is unity, all the ripple voltage across D_2 appears at the output. The ripple voltage that results from variations in the collector is negligible as predicted by Eq. (11-21). Hence the ripple component e_o of e_{OUT} is given by

$$e_o = \frac{r_d}{r_d + R} e_i$$

where e_i is the peak-to-peak ripple of e_{IN} in Fig. 11-19b. The voltage e_i is 3.5 V peak to peak, and, if we assume r_d, the dynamic resistance of the zener diode, to be 10 Ω, e_o becomes

$$e_o = \frac{10}{140 + 10} (3.5) = 0.23 \text{ V}$$

The ripple factor r becomes

$$r = \frac{0.23}{12} = 0.019 = 1.9 \text{ per cent}$$

PROBLEMS

11.1 Verify Eq. (11-2) by performing the integration.

11.2 (a) Find the ripple as given by the definition in Eq. (11-8) for a half-wave rectified sinusoid such as that given in Fig. 11-1b.
(b) How does this value compare with the alternative definition given in Eq. (11-9)?

11.3 (a) Calculate the peak-to-peak ripple voltage in the power supply voltage $v_0(t)$ for the circuit in Fig. P11-3.
(b) What is the value of the dc current through the 500-Ω load?
(c) What is the per cent ripple factor r for this circuit?

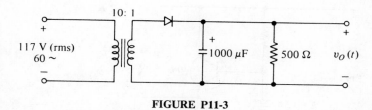

FIGURE P11-3

11.4 Design a power supply of the type shown in Fig. P11-4 to operate the load resistor R_L of 400 Ω with an approximate average output voltage of 30 V and a ripple factor r of not more than 10 per cent. Obtain the value for the capacitor C in the circuit as well as the amount of current that is likely to flow from the 117-V supply.

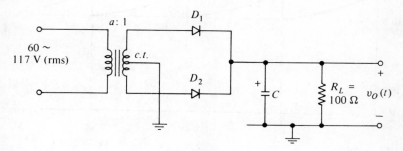

FIGURE P11-4

11.5 Repeat Problem 11.4 employing four diodes for the rectifier portion of the circuit to obtain a bridge rectifier.

11.6 A power source of 80 V dc with 6-V ripple is obtainable in a large analog computer. The source impedance is negligible. A designer wishes to establish a 10-V dc reference for experiment by using a single zener diode. The maximum current that he wishes to drain is 1 mA. What per cent ripple r can he expect on his reference voltage if the dynamic resistance r_d of the zener diode is 18 Ω?

11.7 Design a power supply of the type shown in Fig. 11-18 to deliver 250 mA at 20 V. Use the same technique as was used in Example 11-4.

11.8 Repeat Problem 11.7 using a full-wave bridge rectifier rather than the single diode.

11.9 Repeat Problem 11.7 by using a full-wave rectifier employing two diodes and a center-tapped transformer for the rectifier portion of the circuit.

11.10 Determine R in Fig. P11-10 such that the battery E is being charged with a current of 10 mA.

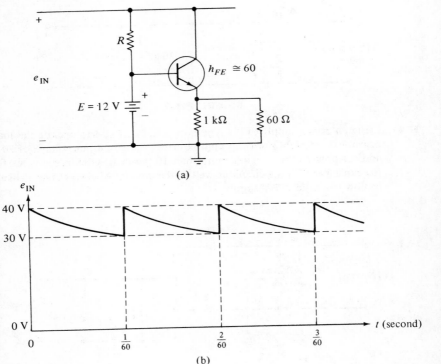

FIGURE P11-10

12

FEEDBACK AMPLIFIERS AND OSCILLATORS

A feedback amplifier may be defined as an amplifier in which the input signal is derived partly from the signal source (the excitation) and partly from the amplifier output.

There are many advantages from the use of feedback in electronic amplifiers. Many of the properties of amplifiers can become improved by careful use of feedback. Feedback can also cause circuits to become unstable, resulting in oscillations. This may be undesirable in some cases, while in others, these circuits can be used as oscillators for signal sources.

12.1 Basic Feedback Theory

To describe feedback mathematically, block diagrams are used. There are two basic elements in a block diagram: the *block* and the *summing junction*. These two elements are shown in Fig. 12-1. The block contains input and output arrows representing the signals, and the interior of the rectangle contains a description of the mathematical operation to be performed on the input to yield the output. In Fig. 12-1a, the output v_o is given by

$$v_o = A_v v_i \tag{12-1}$$

The summing junction contains three signals and represents the sum of two signals. In Fig. 12-1b, the voltage v_i is given by

$$v_i = v_s + v_f \tag{12-2}$$

The connection of these two or more elements constitutes a *system*. In our case, we shall deal with *electronic systems*.

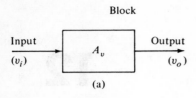

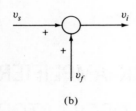

FIGURE 12-1 The two basic elements of a block diagram. (a) The operating block. (b) The summing junction.

A *system block diagram* with *feedback* is shown in Fig. 12-2. This system is comprised of two blocks and a summing junction and is usually called a *basic feedback system*. This block diagram is a mathematical description of a physical system or circuit. The block A_v may represent an amplifier and block β a feedback network, but this need not be so. The block A_v is called

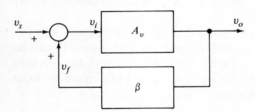

FIGURE 12-2 A block diagram of a system with feedback.

the *forward gain* and is given by

$$A_v = \frac{v_o}{v_i} \tag{12-3}$$

The block β is called the *feedback gain* and is given by

$$\beta = \frac{v_f}{v_o} \tag{12-4}$$

The overall gain A_f is the ratio of v_o to v_s and is sometimes referred to as the *closed loop gain*:

$$A_{v_f} = \frac{v_o}{v_s}$$

The input voltage v_i to block A_v is the sum of the feedback voltage v_f and the signal voltage v_s, and is given by

$$v_i = v_f + v_s \tag{12-5}$$

Substituting the value of v_f from Eq. (12-4), Eq. (12-5) becomes

$$v_i = \beta v_o + v_s \tag{12-6}$$

Eliminating v_i by use of Eq. (12-3), and solving for the overall gain A_{v_f}, we obtain

$$A_{v_f} = \frac{v_o}{v_s} = \frac{A_v}{1 - A_v \beta} \tag{12-7}$$

Example 12-1: The emitter follower circuit shown in Fig. 12-3 is biased by a previous stage. Obtain the small-signal voltage gain of the emitter follower by use of a block diagram.

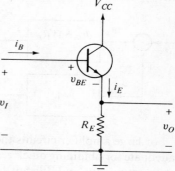

FIGURE 12-3 Emitter follower of Ex. 12-1.

Solution:

We recognize the presence of a summing process in the following statement of Kirchhoff's voltage law:

$$v_i = v_{be} + v_o$$

or

$$v_{be} = v_i - v_o \tag{12-8}$$

The base-to-emitter voltage v_{be} is also given by

$$v_{be} \simeq i_b h_{ie} \tag{12-9}$$

and the output voltage v_o is given by

$$\begin{aligned} v_o &= i_e R_E \\ &= (h_{fe} + 1) R_E i_b \end{aligned} \tag{12-10}$$

Combining Eqs. (12-9) and (12-10), we obtain

$$v_o = \frac{(h_{fe} + 1) R_E}{h_{ie}} v_{be} \tag{12-11}$$

Using Eqs. (12-8) and (12-11) we can now construct the block diagram shown in Fig. 12-4. Using Eq. (12-7) we can calculate the overall gain to be

$$A_{v_f} = \frac{v_o}{v_i} = \frac{A_v}{1 - \beta A_v}$$

$$= \frac{\dfrac{(h_{fe} + 1)R_E}{h_{ie}}}{1 + \dfrac{(h_{fe} + 1)R_E}{h_{ie}}} \quad (12\text{-}12)$$

$$= \frac{(h_{fe} + 1)R_E}{(h_{fe} + 1)R_E + h_{ie}}$$

which yields the correct result for the voltage gain of an emitter follower.

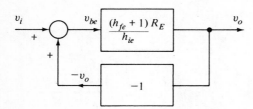

FIGURE 12-4 Block diagram for the emitter follower of Ex. 12-1.

For large amplifier circuits the technique used in Example 12-1 is usually inadequate for obtaining quick results. The equations describing the amplifier may be numerous and difficult to obtain. A more suitable approach can be developed using a two-port equivalent for an amplifier and then connecting a two-port equivalent for the feedback circuit to this amplifier.* Such an interconnection is shown in Fig. 12-5. The two-port equivalent of the amplifier may be obtained from our knowledge of its input impedance, output impedance, and voltage gain.

The diagram of Fig. 12-5 represents a rather idealized situation. However, if the quantities Z_{in}, A_v, Z_{out}, and β are carefully defined, the theory that we develop can be used to approximate a large number of practical examples.

The circuit of Fig. 12-5 is usually called a *voltage feedback* system, since the signal that is fed back from the output to the input is a voltage.

In Fig. 12-5, A_v represents the open circuit gain of the amplifier without feedback and is called the *forward gain* or the *open loop gain*. It can be found by removing V_s, leaving an open circuit, and removing Z_{ext}, the external load. A voltage V_i is applied at terminals 1, 1', and the output voltage is measured at terminals 2, 2'. The open loop gain A_v then becomes

$$A_v = \frac{V_o}{V_i} \quad (I_l = 0) \quad (12\text{-}13)$$

---

*The treatment of feedback amplifiers is treated in a brief conventional manner. For a more comprehensive treatment, see Ref. [1].

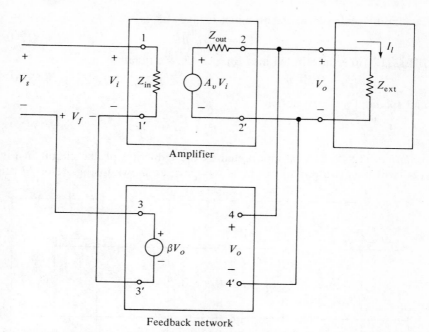

FIGURE 12-5 An equivalent circuit of an amplifier with voltage feedback.

The output impedance Z_{out} of the amplifier can be found by removing Z_{ext} and V_s, shorting terminals 1 and 1′, and measuring the impedance looking into terminals 2 and 2′. It is interesting to note that in a physical interconnection of two two-ports, as shown in Fig. 12-5, the loading effects of the feedback network have been accounted for in the manner by which the definitions were made. That is, if the rules used to define A_v and Z_{out} are used on an actual circuit, the equivalent circuit that results will be that of Fig. 12-5.

The *input impedance* Z_{in} of the amplifier is the impedance looking in at terminals 1, 1′, with V_s removed from the circuit. The *feedback factor* β of Fig. 12-5 defines the amount of the voltage that is impressed at terminals 1, 1′ due to V_o if V_s is set to zero. (Short V_s to set it equal to zero.)

$$\beta = \left.\frac{V_i}{V_o}\right|_{V_s=0} \tag{12-14}$$

Notice that β must be independent of the value of the external load Z_{ext}. A change in Z_{ext} must not produce a change in β.

To analyze the action of the circuit of Fig. 12-5, we write an expression for the output as

$$V_o = A_v V_i - I_l Z_{out} \tag{12-15}$$

The input voltage V_i to the amplifier is given as

$$V_i = V_s + \beta V_o \qquad (12\text{-}16)$$

If Eq. (12-16) is substituted into Eq. (12-15), we obtain

$$V_o = A_v(V_s + \beta V_o) - I_l Z_{out} \qquad (12\text{-}17)$$

and solving for V_o we obtain

$$V_o = \frac{A_v}{1 - A_v\beta} V_s - \frac{Z_{out}}{1 - A_v\beta} I_l \qquad (12\text{-}18)$$

Equation (12-18) can be interpreted as the reduction of the circuit in Fig. 12-5 with two two-ports to a single two-port, as shown in Fig. 12-6. At this

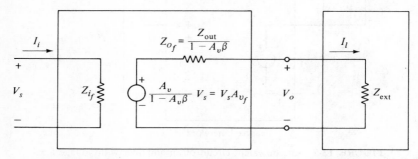

FIGURE 12-6 The reduced equivalent circuit of an amplifier with feedback.

point it is convenient to define the *closed loop gain* A_{v_f} as being the voltage gain of an amplifier with feedback, with the external load current set at zero. Mathematically, the closed loop gain is defined as

$$A_{v_f} = \left.\frac{V_o}{V_s}\right|_{I_l=0} = \frac{A_v}{1 - A_v\beta} \qquad (12\text{-}19)$$

It is interesting to note that this is the same as Eq. (12-7), which was obtained from the block diagram representation of feedback.

The second term in Eq. (12-18), which describes the effect of the load current I_l on the output voltage, can be interpreted as a change of output impedance by comparing it with Eq. (12-15), which describes the action of a network without feedback. Thus we relate the output impedance of the network with feedback Z_{o_f} to the output impedance of the amplifier Z_{out} as follows:

$$Z_{o_f} = \frac{Z_{out}}{1 - A_v\beta} \qquad (12\text{-}20)$$

To obtain the input impedance in Fig. 12-6, which is the input impedance of the amplifier under the effect of feedback Z_{i_f}, we must find the ratio of the input voltage V_s to the input current I_i. To do this we write the equation for the input voltage V_s in Fig. 12-5 as

$$V_i = V_s + \beta V_o \qquad (12\text{-}21)$$

Solving for V_s from Eq. (12-21), we obtain

$$V_s = V_i - \beta V_o$$
$$= V_i(1 - A_v\beta) \qquad \text{(with } I_t = 0\text{)}$$

Finally, the input impedance Z_{i_f} in terms of Z_i is

$$Z_{i_f} = \frac{V_s}{I_i} = \frac{V_i}{I_i}(1 - A_v\beta) \qquad (12\text{-}23)$$
$$= Z_i(1 - A_v\beta)$$

At this point it is useful to define *negative feedback* as being the condition where the magnitude $1 - A_v\beta$ is greater than unity:

$$|1 - A_v\beta| > 1 \qquad (12\text{-}24)$$

The magnitude in Eq. (12-24) implies that A or β, or both, can be functions of frequency and hence functions of $j\omega$. Therefore, it is conceivable that an amplifier may be a negative-feedback amplifier for some frequencies, yet not for others. When the condition in Eq. (12-24) *is not met* we say we have *positive feedback* occurring. Thus, from Eqs. (12-19), (12-20), and (12-23), for a negative feedback amplifier, we can state the following:

1. The overall voltage gain, or closed loop gain, is decreased in magnitude by the factor $|1 - A_v\beta|$.
2. The output impedance is decreased in magnitude by $|1 - A_v\beta|$.
3. The input impedance is increased in magnitude by the factor $|1 - A_v\beta|$.

It is difficult to consider all the consequences of Eq. (12-24) at this point, so we shall consider the case where A_v and β are both real quantities. This case would arrive in practice when we are considering an amplifier in the mid-band frequency range where the gain A_v is either a positive or negative quantity. The real β can be easily obtained by using resistive networks. For the case of real A_v and real β, Eq. (12-24) can be written as

$$1 - A_v\beta > 1 \qquad (12\text{-}25)$$
$$A_v\beta < 0 \qquad (12\text{-}26)$$

This means that for a negative-feedback amplifier we require the product $A_v\beta$ to be negative.

12.2 Circuits with Negative Voltage Feedback

Since the properties of low output impedance and high input impedance are generally desirable, voltage feedback is used quite frequently. The amplifier shown in Fig. 12-7 is a very common circuit employing negative feedback. To show that this is a feedback circuit we should establish the basic elements of the circuit: the summing junction, the amplifier block, and the feedback network. The summing at the input is apparent if the voltages are designated, as shown in Fig. 12-7:

$$v_i = v_s + v_f \tag{12-27}$$

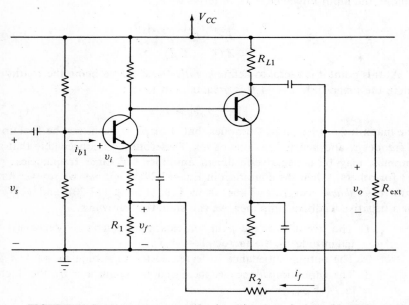

FIGURE 12-7 A two-stage amplifier in which the principles of voltage feedback can be applied.

The feedback network can be thought of as a simple voltage divider network:

$$\beta = \frac{-R_1}{R_1 + R_2} = \frac{v_f}{v_o} \tag{12-28}$$

Equation (12-28) will be approximately true as long as several conditions are met:

1. To make β independent of R_{ext}, R_2 should be much larger than R_{L1} in parallel with R_{ext}:

$$R_2 \gg R_{L1} \| R_{ext} \tag{12-29}$$

2. To make v_f due primarily to the output voltage and not to the small-signal emitter current arising from i_{b1}, the following condition must also be met:

$$i_f R_1 > h_{fe} i_{b1} R_1$$

or

$$i_f > h_{fe}i_{b1} \tag{12-30}$$

where h_{fe} is the short circuit gain of the first transistor in Fig. 12-7. This would give us a perfect summing junction.

It is easy to see that condition (2) is not met simultaneously with condition (1). If condition (2) is not met, then an additional feedback phenomenon occurs in the input, which is similar to that for the emitter follower. This increases the complexity of the problem. A rather simple solution is to try to meet conditions (1) and (2) as well as possible, do the calculations assuming they are met, and correct R_2 experimentally to obtain the desired gain.

The external load resistor R_{ext} can be treated as part of the amplifier circuit. This will decrease both the gain and the output impedance. The open loop gain A_v for an amplifier of this type is quite high, in the order of 10,000 or 20,000. This is the ratio of v_o to v_i, with v_f equal to zero. The voltage v_f can be set at zero by removing R_2. Using Eq. (12-19), the overall gain of the amplifier becomes

$$A_{v_f} = \frac{v_o}{v_s} = \frac{A_v}{1 - A_v\beta} = \frac{A_v}{1 + \dfrac{A_vR_1}{R_1 + R_2}} \tag{12-31}$$

Equation (12-31) can be simplified if A_v is very large, or

$$A_v\frac{R_1}{R_1 + R_2} > 1 \tag{12-32}$$

Then Eq. (12-31) becomes

$$\frac{v_o}{v_s} \simeq \frac{R_1 + R_2}{R_1} = 1 + \frac{R_2}{R_1} \tag{12-33}$$

Equation (12-33) can be further simplified if R_2 is greater than R_1, as it should be, if gain greater than unity is to be obtained:

$$\frac{v_o}{v_s} \simeq \frac{R_2}{R_1} \quad \text{if } R_2 > R_1 \tag{12-34}$$

It is interesting to note that we do have a negative-feedback circuit, since the product of $A_v\beta$ is negative. The voltage gain A_v is positive, since two CE stages are used and β is negative, as was shown in Eq. (12-28).

The result of Eq. (12-34) agrees quite well with practical circuits, even if both conditions (1) and (2) are not met. This circuit finds much use for filter, equalizer and preamplifier design by using an impedance Z_2, which varies with frequency in place of R_2. The resulting gain is approximately proportional to the impedance used. For example, if a parallel R-C circuit is used, a flat low-frequency response is achieved with a decrease at the high frequencies because of the shorting effect of the capacitance.

12.3 Additional Advantages of Negative Feedback

In section 12-2 we saw the effect of negative feedback on the input impedance, output impedance, and voltage gain. The effects obtained are generally desirable in the design of amplifiers. The use of negative feedback in amplifiers also has a tendency to decrease noise and distortion, maintain the operation due to parameter changes in the circuit, and improve the frequency response. In this section we shall discuss each of these effects in turn.

First, to examine the effect of distortion or noise we will consider the block diagram shown in Fig. 12-8a. Let us assume that noise or distortion arises somewhere in the middle of the amplifier and produces a noise voltage N_s. The amplification that occurs before this noise is designated by the block A_1, and A_2 represents the amplification that occurs after N_s is injected into the circuit. β represents the feedback network.

From the block diagram of Fig. 12-8a we can write the expression for the output voltage V_o as follows:

$$V_o = A_2(N_s + V_i A_1) \tag{12-35}$$

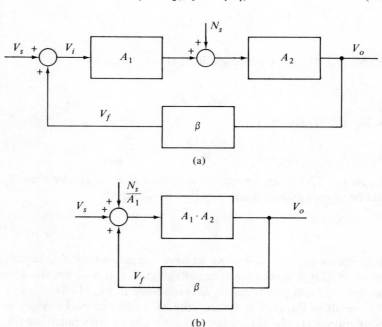

(a) Block diagram of a feedback amplifier with noise.

(b) The simplified block diagram with the noise referred to the input.

FIGURE 12-8 Feedback block diagram with a noise input.

At the summing junction we can express V_i as

$$V_i = \beta_o V_o + V_s \qquad (12\text{-}36)$$

Substituting Eq. (12-36) into (12-35), we obtain

$$\begin{aligned}V_o &= A_2[N_s + (\beta V_o + V_s)A_1] \\ &= A_2 N_s + A_1 A_2 \beta V_o + A_1 A V_s\end{aligned} \qquad (12\text{-}37)$$

Solving for V_o in Eq. (12-37), we obtain

$$V_o = \frac{A_2 N_s}{1 - A_1 A_2 \beta} + \frac{A_1 A_2 V_s}{1 - A_1 A_2 \beta} \qquad (12\text{-}38)$$

If we now let A equal $A_1 A_2$, which is the overall open loop gain of the amplifier, Eq. (12-37) can be rewritten as follows:

$$V_o = \frac{N_s}{A_1} \frac{A}{1 - A\beta} + \frac{A}{1 - A\beta} V_s \qquad (12\text{-}39)$$

Equation (12-39) can now be described by the block diagram shown in Fig. 12-8b. It is the same as the basic feedback system except that there are two inputs, V_s and N_s/A_1. The quantity N_s/A_1 is the noise voltage referred to the input. If it were not for the feedback due to β, the output voltage due to N_s would simply be the noise referred to the input multiplied by the open loop gain A. However, from Eq. (12-39), it is apparent that this noise referred to the input is reduced in magnitude by the presence of the feedback by the quantity $1 - A\beta$, the magnitude of this quantity being greater than unity for negative-feedback systems.

Variations in the transistor amplifier components or the transistor parameters due to aging, temperature, component replacement, etc., would result in a change in the gain of the amplifier. The per cent change of the amplifier gain with feedback can be expressed as

$$\frac{dA_f}{A_f} \simeq \frac{\Delta A_f}{A_f} = \frac{\text{change in amplification gain}}{\text{value of the gain}} \qquad (12\text{-}40)$$

By differentiating the expression for the closed loop gain it is easy to show that the per cent change in amplifier gain with feedback is related to the per cent change in the amplifier without feedback.

$$\frac{dA_f}{A_f} = \frac{1}{1 - \beta A} \frac{dA}{A} \qquad (12\text{-}41)$$

Equation (12-41) shows that the per cent change of the amplification in a feedback amplifier is reduced in magnitude by the factor $|1 - \beta A|$, for a negative-feedback system.

The frequency response of the gain may also be improved by the use of negative feedback. Let us assume that we have an amplifier of the type shown in Fig. 12-7, where the frequency response of A_v can be expressed approximately as

$$A_v(s) = \frac{K}{1 + \dfrac{s}{\omega_H}} = \frac{K\omega_H}{s + \omega_H} \tag{12-42}$$

and let us assume that β is a *real negative* number so that the product $A_v\beta$ is negative in the mid-band region, giving us negative feedback.

The closed loop gain A_{v_f} for this amplifier is given by

$$\begin{aligned}A_{v_f} &= \frac{A_v}{1 - \beta A_v} \\ &= \frac{\dfrac{K\omega_H}{s + \omega_H}}{1 - \beta K \dfrac{\omega_H}{s + \omega_H}} \\ &= \frac{K\omega_H}{s + \omega_H(1 - \beta K)} \\ &= \frac{\dfrac{K}{1 - \beta K}}{1 + \dfrac{s}{\omega_H(1 - \beta K)}}\end{aligned} \tag{12-43}$$

Notice that the mid-band gain is reduced by the factor $1 - \beta k$, which is a positive number greater than unity, and the frequency cutoff is increased by the same amount of $1 - \beta K$. A Bode plot of the two gains is shown in Fig. 12-9.

The same procedure can be applied to find the low-frequency response for a feedback amplifier. The result is that the introduction of feedback decreases the lower cutoff frequency by the factor $1/(1 - \beta K)$.

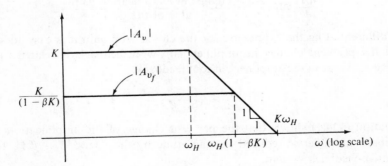

FIGURE 12-9 A Bode diagram illustrating the effect of voltage feedback on the voltage-gain A_v, as well as the effect on the upper cutoff frequency.

12.4 Current-Feedback Amplifiers

A current-feedback amplifier is one in which the input current is partly derived from the output current. A block diagram that mathematically describes such a system is shown in Fig. 12-10a. The system is identical to a voltage-feedback system, except that the signals are currents rather than voltages. The block A_i represents the open loop current gain and β represents the current-feedback gain. Since the block diagram and hence the equa-

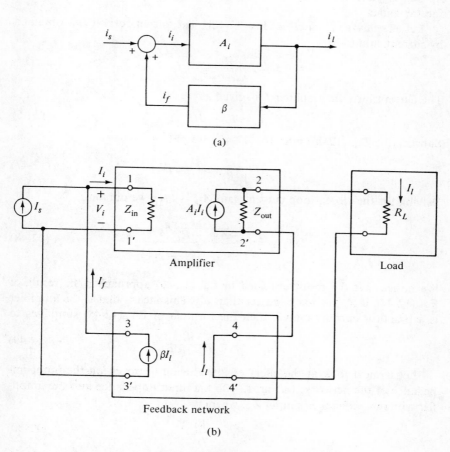

(a) Block diagram of current feedback system
(b) A two-port implementation of the current feedback system

FIGURE 12-10 Current feedback diagrams.

tions that describe a current-feedback system are the same as for the voltage-feedback system, it is obvious that the closed loop current gain A_{i_f} is given by

$$A_{i_f} = \frac{i_l}{i_s} = \frac{A_i}{1 - \beta A_i} \tag{12-44}$$

A two-port equivalent circuit of an amplifier with current feedback is shown in Fig. 12-10b. As with voltage feedback, this represents a rather idealized situation that approximates the practical case. The manner in which Z_{in}, A_i, and Z_{out} can be measured or calculated in practice is left as an exercise for the reader.

Let us analyze the circuit of Fig. 12-10b. The output current I_l is obtained by current gain to be

$$I_l = \frac{Z_{out}}{Z_{out} + R_L} A_i I_i \tag{12-45}$$

The current into the amplifier I_i is given as

$$I_i = I_s + \beta I_l \tag{12-46}$$

Substituting Eq. (12-46) into Eq. (12-45), we obtain

$$I_l = \frac{Z_{out}}{Z_{out} + R_L} A_i (I_s + \beta I_l) \tag{12-47}$$

Solving for the closed loop current gain $A_{i_f} = I_l/I_s$, we obtain

$$A_{i_f} = \frac{I_l}{I_s} = \frac{\dfrac{Z_{out}}{Z_{out} + R_L} A_i}{1 - \dfrac{Z_{out}}{Z_{out} + R_L} \beta A_i} \tag{12-48}$$

We notice that the result obtained in Eq. (12-48) approaches the result of Eq. (12-44), if Z_{out} is much greater than R_L. This means that if the amplifier is a practical current source, then the formula of Eq. (12-48) simplifies to

$$A_{i_f} = \frac{I_l}{I_s} = \frac{A_i}{1 - \beta A_i} \tag{12-49}$$

Let us next look at the effect of introducing feedback on the input impedance of the network. In Fig. 12-10b the input impedance into the amplifier without feedback is simply Z_{in}, where

$$Z_{in} = \frac{V_i}{I_i} \tag{12-50}$$

With the introduction of feedback, the input impedance of the whole system is expressed in the ratio

$$Z_{i_f} = \frac{V_i}{I_s} \tag{12-51}$$

Since I_s is given by

$$I_s = I_i - \beta I_i$$
$$= I_i\left(1 - \frac{Z_{out}}{Z_{out} + R_L}\beta A_i\right) \quad (12\text{-}52)$$

and the substitution of Eq. (12-52) into Eq. (12-51) yields

$$Z_{i_f} = \frac{Z_{in}}{1 - \frac{Z_{out}}{Z_{out} + R_L}\beta A_i} \quad (12\text{-}53)$$

simplified to

$$Z_{i_f} = \frac{Z_{in}}{1 - \beta A_i} \quad \text{if } Z_{out} > R_L \quad (12\text{-}54)$$

it is easy to see that for a negative-feedback system the magnitude of the input impedance decreases when current feedback is used.

To determine the effect of current feedback on the output impedance as seen by R_L, we find the output impedance by taking the ratio of the open-circuit voltage to the short-circuit current. Thus the output impedance with feedback Z_{out_f} is given by

$$Z_{out_f} = \frac{\text{open circuit voltage at the output}}{\text{short circuit current at the output}} \quad (12\text{-}55)$$

The open-circuit voltage V_{oc} is

$$V_{oc} = A_i I_i Z_{out}$$
$$= A_i I_s Z_{out} \quad (\text{since } \beta i_t = 0) \quad (12\text{-}56)$$

The short-circuit current I_{sc} is

$$I_{sc} = \frac{A_i}{1 - A_i\beta} I_s \quad (12\text{-}57)$$

The ratio of Eqs. (12-56) and (12-57) yields the desired result—the output impedance Z_{out_f}:

$$Z_{out_f} = Z_{out}(1 - A_i\beta) \quad (12\text{-}58)$$

From Eq. (12-58) it is easy to see that the magnitude of the output impedance of the amplifier increases when negative current feedback is used.

Example 12-2: The common emitter circuit with feedback biasing, shown in Fig. 12-11, is driven by a voltage source with an internal impedance of R_S, which is comparable in size to h_{ie} of a transistor.

(a) Use the theory of current feedback to obtain R_{in_f} input impedance into the amplifier.

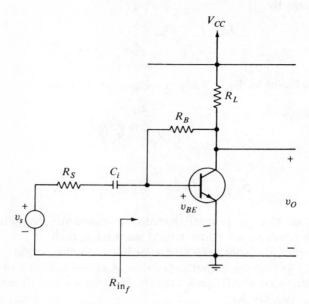

FIGURE 12-11 A CE amplifier with feedback biasing used in example problem 12-2.

(b) Use the following typical values for the circuit components to calculate a value for R_{in_f}.

$$h_{ie} = 1000 \; \Omega$$
$$h_{fe} = 50$$
$$R_L = 5 \; k\Omega$$
$$R_B = 200 \; k\Omega$$
$$R_S = 1 \; k\Omega$$

(c) Calculate the overall voltage gain v_o/v_s.

Solution:

(a) The equivalent circuit for the amplifier is obtained as shown in Fig. 12-12. We recognize the open loop current gain A_i of the amplifier block to be

$$A_i = -h_{fe}$$

To obtain the feedback factor β we return to Fig. 12-10 and see that we must set the driving source i_s equal to zero (in this example it is v_s/R_S) and calculate the factor, β, which determines the amount of input current that flows as

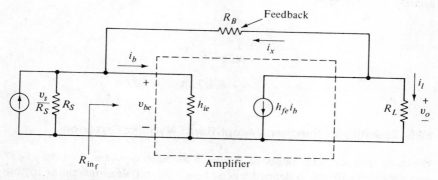

FIGURE 12-12 The equivalent circuit for a feedback biased CE amplifier.

a result of the output current. Thus, the factor β becomes

$$\beta \simeq \frac{R_L}{R_B + R_L} \frac{R_S}{R_S + h_{ie}}$$

The factor β was obtained by using current division twice, first between R_L and R_B and then again between R_S and h_{ie}. Thus R_{in_f} becomes

$$R_{in_f} = \frac{h_{ie}}{1 - \beta A_i}$$

$$= \frac{h_{ie}}{1 + h_{fe} \dfrac{R_L}{R_B + R_L} \dfrac{R_S}{R_S + h_{ie}}}$$

(b)
$$1 - \beta A_i = 1 + (50)\frac{5\text{ k}\Omega}{255\text{ k}\Omega} \frac{1\text{ k}\Omega}{2\text{ k}\Omega}$$

$$\simeq 1 + \tfrac{1}{2} = 1.5$$

$$R_{in_f} = \frac{h_{ie}}{1 - \beta A_i} = \frac{1{,}000}{1.5} \simeq 670\ \Omega$$

(c) To calculate the voltage gain we first observe that the voltage gain v_o/v_{be} remains almost the same if R_B is not in the equivalent circuit. The reason is that i_l is much greater than i_x in Fig. 12-12, and the output voltage remains unaffected. However, the input impedance has dropped owing to the presence of R_B and hence the voltage gain v_{be}/v_s. Using voltage division, we obtain

$$\frac{v_{be}}{v_s} = \frac{R_{in_f}}{R_{in_f} + R_S} = \frac{h_{ie}}{h_{ie} + R_S(1 - \beta A_i)}$$

$$= \frac{1\text{ k}\Omega}{1\text{ k}\Omega + 1\text{ k}\Omega(1.5)} = 0.4$$

Finally, v_o/v_s becomes

$$\frac{v_o}{v_s} = \frac{v_{be}}{v_s}\frac{v_o}{v_{be}}$$

$$= 0.4\frac{-h_{fe}R_L}{h_{ie}}$$

$$= -0.4(250) = -100$$

12.5 Stability of Feedback Amplifiers: Nyquist Criterion

We have already seen negative feedback as defined when $|1 - A\beta| > 1$* and positive feedback as defined when $|1 - A\beta| < 1$. It is interesting to note that the magnitude of the gain, be it voltage or current, decreases under the effect of negative feedback. For this reason, it is often called *degenerate feedback*. Positive feedback, on the other hand, causes an increase in gain and is called *regenerative feedback*.

Positive feedback can be used very effectively to increase the gain but is very seldom used, since it may lead to oscillation in the amplifier. To explain how this can happen we must understand more intensively how a positive-feedback amplifier operates. To illustrate a case, consider the simple block diagram of a feedback system (Fig. 12-2), where A and β are both real and positive and the product $A\beta$ is just greater than unity. This will certainly make it positive feedback since $|1 - A\beta|$ is less than unity. Assume that a small signal v_i is introduced because of some disturbance in the signal. This signal would pass through both A and β and appear as v_f, which would be slightly larger than v_i, causing v_i to be reinforced or regenerated. Hence this signal would also regenerate itself in the same manner, resulting in a large buildup of this signal. This would probably result in oscillation, since the size of the built-up voltage would be limited by the power supply, the maximum voltage without clipping, etc. We would then say that the amplifier is unstable.

Nyquist has devised a stability criterion that may be used to determine whether a network is stable or not. Nyquist's criterion involves plotting the product $A\beta$ in the complex plane of $A\beta$—that is, plotting $A\beta$ for all frequencies on an Argan diagram, where the X-axis is the real value of $A\beta$ and the Y-axis is the imaginary value. $A\beta$ must be plotted for all frequencies from 0 to $+\infty$. When this is done, Nyquist's criterion may be stated as follows: *The amplifier is unstable if the resulting curve encloses the $+1 + j0$ point, and stable if the curve does not enclose this point.* This point is often called the *critical point*.

This criterion does not state that negative-feedback systems are stable and

---

*A can be either A_i or A_v. The omission of the subscript represents either voltage or current gain.

positive-feedback systems are unstable. To understand the interrelationship between these two ideas we should consider the locus of $|1 - A\beta|$ in the $A\beta$ plane. If all the points that satisfy the condition $|1 - A\beta| < 1$ were plotted in the $A\beta$ plane, it would result in the shaded circular region shown in Fig. 12-13. Notice that the critical point lies in the center of the region of positive feedback. Thus it is possible for a curve of $A\beta$ to pass through the positive-feedback region without enclosing the $+1 + j0$ point. Such an amplifier would be stable.

If a plot of $A\beta$ were made for a simple stage amplifier, discussed in Example 12-2, the resulting curve would be of the type shown in Fig. 12-14. We notice that the amplifier is stable and operates as a negative feedback amplifier for all frequencies.

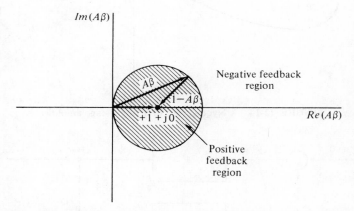

FIGURE 12-13 The complex plane of $A\beta$ illustrating the positive and negative feedback regions.

12.6 Sinusoidal Oscillators and the Barkhausen Stability Criterion

As was stated in the introduction, circuits that oscillate at a single frequency are useful as signal sources or signal generators. It is for this reason that we now study the unstable feedback amplifier. Consider the block diagram of a feedback amplifier (voltage or current) shown in Fig. 12-15. The symbol χ is used to represent either a voltage or current. The closed loop gain is given by

$$\frac{\chi_o}{\chi_s} = \frac{A}{1 - A\beta} \qquad (12\text{-}59)$$

It is easy to see that if $A\beta$ is precisely unity, then the gain becomes infinity,

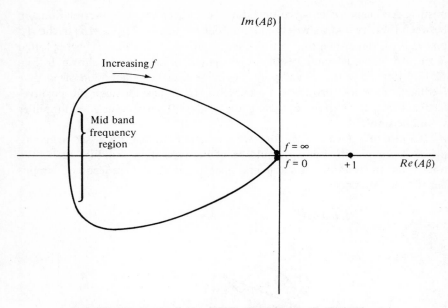

FIGURE 12-14 A Nyquist plot of a single-stage amplifier.

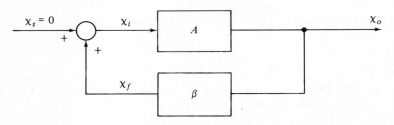

FIGURE 12-15 A generalized block diagram of a feedback system.

which may be interpreted as a device that delivers an output χ_o even when the input χ_s is zero.

The Barkhausen criterion states that *a sinusoidal oscillator will oscillate at a frequency in which the total phase shift of A and β is precisely zero (or multiples of 2π)*. This means that a signal χ_i caused by a disturbance will tend to be regenerated by an *in-phase* signal χ_f. Barkhausen's criterion goes on to say: *Oscillations will not be sustained at this frequency if the value of $A\beta$ at this frequency is less than unity*. In other words, the Barkhausen criterion implies that in order to obtain sustained oscillations in a feedback amplifier the magnitude of $A\beta$ should be at least unity and the phase of $A\beta$ is zero, or multiples of 2π. If this is related to the Nyquist criterion, it means that the locus of $A\beta$ in the $A\beta$ plane must encircle the critical point to have sustained oscillations.

12.7 Practical Oscillators

Practical oscillator circuits employing transistors are generally very easy to analyze and design. Figure 12-16a illustrates a simplified diagram of an oscillator using a single transistor. This simple oscillator uses a transistor to provide the forward-current gain A_i and a feedback network, which includes the biasing network, to provide the proper phase shift. Figure 12-16b shows a block diagram of the circuit. Notice that A_i is a negative real number. To satisfy the Barkhausen criterion the feedback network must supply an additional 180° phase shift at some frequency ω_o so that the product of $A_i\beta$ at ω_o can be at least unity.

Figure 12-16c illustrates how $A_i\beta$ can be calculated for a particular network. Since the input impedance of the transistor in Fig. 12-16a is h_{ie}, breaking the loop as shown in Fig. 12-16b must include the loading effects of the transistor on the feedback network. Thus h_{ie} must be added to the end of the feedback network when the open loop gain $A_i\beta$ is to be calculated in practical transistor oscillator circuits of this type.

One of the best examples in which this method can be used to determine the frequency of oscillation, and hence the design equations, is the phase-shift oscillator shown in Fig. 12-17a. The transistor T_1 is biased as usual in the CE configuration employing H-type biasing with the feedback network consisting of the network made of an interconnection of capacitors and resistors. The equivalent circuit for the phase-shift oscillator with the feedback loop opened, as described previously, is shown in Fig. 12-17b. Notice that R_L is expressed in terms of the resistor R:

$$R_L = kR \tag{12-60}$$

It is also important to notice that the value of R_F is such that the combination of R_F, R_A, R_B, and h_{ie} in Fig. 12-17b is equal to R:

$$R = R_F + (R_A || R_B || h_{ie}) \tag{12-61}$$

Figure 12-17c shows an equivalent circuit of the phase-shift oscillator with a Thévenin equivalent source. To obtain the design equations for this oscillator we must obtain the open loop gain I_f/I_b and set this equal to unity. This current gain can be found by solving the following simultaneous equations, which can be obtained from the method of mesh analysis:

$$-kRh_{fe}I_b = \left[R(1+k) + \frac{1}{j\omega C}\right]I_1 - RI_2 \tag{12-62}$$

$$0 = -RI_1 + \left(2R + \frac{1}{j\omega C}\right)I_2 - RI_3 \tag{12-63}$$

$$0 = -RI_2 + \left(2R + \frac{1}{j\omega C}\right)I_3 \tag{12-64}$$

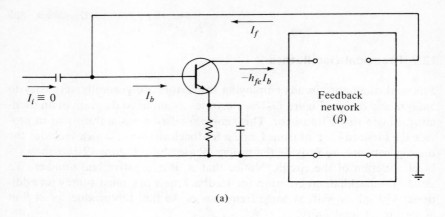

(a)

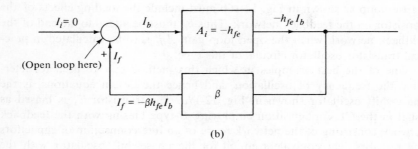

(b)

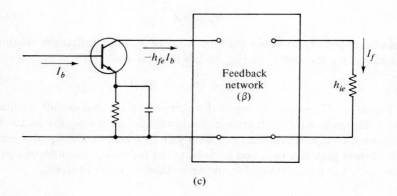

(c)

(a) A typical feedback circuit employed in oscillator design.
(b) A block diagram representation of the oscillator circuit.
(c) The open-loop circuit of the oscillator.

FIGURE 12-16 Basic oscillator circuit and equivalent circuits.

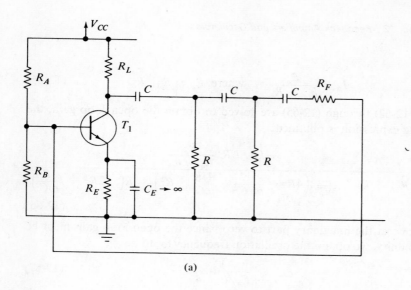

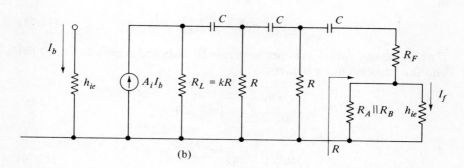

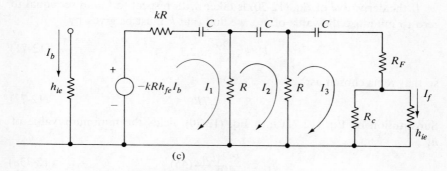

(a) A typical RC phase-shift oscillator.
(b) The equivalent circuit of the RC phase-shift oscillator.
(c) The open-loop circuit with mesh current directions indicated.

FIGURE 12-17 Phase-shift oscillator and equivalent circuits.

332   Chap. 12 / Feedback Amplifiers and Oscillators

and

$$I_f = \frac{R_c}{h_{ie} + R_c} \quad \text{where } R_c = R_A \| R_B \tag{12-65}$$

If Eqs. (12-62) through (12-65) are solved to obtain the open loop gain, the following expression is obtained:

$$\frac{I_f}{I_b} = \frac{-kh_{fe}I_bR^3 \dfrac{R_c}{R_c + h_{ie}}}{\left[R(1+k) - \dfrac{j}{\omega C}\right]\left(4R^2 - \dfrac{1}{\omega^2 C^2} - \dfrac{4jR}{\omega C} - R^2\right) + R\left(-2R^2 + j\dfrac{R}{\omega C}\right)} \tag{12-66}$$

If we now set the imaginary part to zero, since the open loop gain must be equal to unity, we obtain the oscillation frequency ω_o to be

$$\omega_o = \frac{1}{RC\sqrt{6 + 4k}} \tag{12-67}$$

or

$$(\omega_o CR)^2 = \frac{1}{6 + 4k} \tag{12-68}$$

If we set the real part of the open loop gain to unity and substitute Eq. (12-68) into this expression, we obtain

$$\frac{kh_{fe} \dfrac{R_c}{R_c + h_{ie}}}{4k^2 + 23k + 29} = 1 \tag{12-69}$$

or

$$\frac{R_c h_{fe}}{R_c + h_{ie}} = 4k + 23 + \frac{29}{k} \tag{12-70}$$

If the derivative of Eq. (12-70) is taken with respect to k and set equal to zero to minimize the value of h_{fe}, we find that k must be given by

$$k \simeq \sqrt{\frac{29}{4}} \simeq 2.7 \tag{12-71}$$

so that R_L is chosen as

$$R_L = 2.7R \tag{12-72}$$

Substitution of Eq. (12-71) into Eq. (12-70) yields the minimum value of h_{fe}:

$$h_{fe} \simeq \frac{R_c + h_{ie}}{R_c}(45) \tag{12-73a}$$

This procedure completes the analysis of the phase-shift oscillator. To design such an oscillator we should bear in mind the following assumptions and results given in Eqs. (12-61), (12-67), (12-71), (12-72), and (12-73a). They are repeated here in more convenient forms as a handy summary for design purposes:

$$R_F = R + (R_A || R_B || h_{ie}) \tag{12-61}$$

$$\omega_o = \frac{1}{RC\sqrt{6 + 4k}} \quad \text{where } k = 2.7 \tag{12-67}$$

$$R_L = 2.7R$$

$$h_{fe} \geq \frac{R_c + h_{ie}}{R_c}(45) \quad \text{where } R_c = R_A || R_B \tag{12-73b}$$

The expression in Eq. 12-73b has been altered from the original Eq. (12-73a) with a greater-than sign, since, according to the Barkhausen criterion, a circuit will oscillate if the open loop gain is greater than unity, provided the phase shift is zero. This will certainly be the case if h_{fe} is chosen larger than the value in Eq. (12-73a).

Another oscillator circuit in which the same procedure can be employed is the Colpitts oscillator, shown in Fig. 12-18a. The equivalent circuit necessary to find the open loop gain is shown in Fig. 12-18b. The design equations are found to be

$$\omega_o = \frac{1}{\sqrt{LC}} \quad \text{where } C = \frac{C_1 C_2}{C_1 + C_2} \tag{12-74}$$

$$h_{fe} \geq \frac{R_E + h_{ie}}{R_E} \frac{C_2}{C} \simeq \frac{R_E + h_{ie}}{R_E} \frac{C_2}{C_1} \tag{12-75}$$

It is interesting to note from Eq. (12-75) that C_2 will generally be made much larger than C_1, and so the value of C becomes essentially the value of C_1. In this way, ω_o can be varied over a large range of frequencies by using a variable capacitor for C_1.

Another interesting oscillator circuit, the Wein Bridge oscillator, is shown symbolically in Fig. 12-19a. It is an excellent example of a circuit designed according to voltage feedback concepts. It consists of an ideal voltage amplifier having infinite input impedance, zero output impedance, and a voltage gain of A, as well as a feedback network consisting of two resistors and two capacitors, connected as shown.

The feedback factor β can be calculated using voltage division as follows:

$$\beta = \frac{\dfrac{R}{1 + j\omega CR}}{R + \dfrac{1}{j\omega C} + \dfrac{R}{1 + j\omega CR}}$$

$$= \frac{RC}{3RC + j\omega C^2 R^2 + \dfrac{1}{j\omega C}} \tag{12-76}$$

If a relatively low frequency oscillator is to be designed, for audio purposes

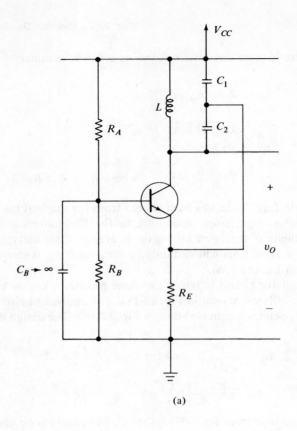

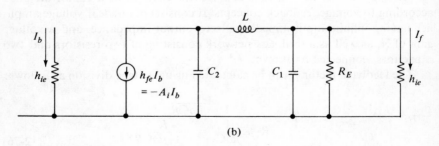

(a) A transistor Colpitts oscillator.
(b) The open-loop equivalent circuit of the Colpitts oscillator.

FIGURE 12-18 Colpitts oscillator and equivalent circuit.

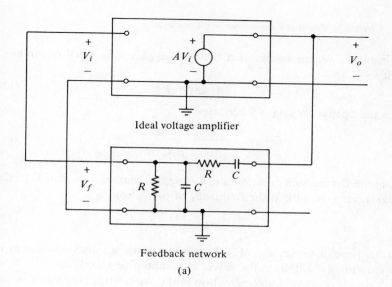

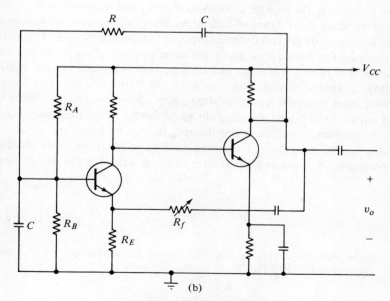

(a) Basic Wien bridge oscillator circuit.
(b) Transistor Wien bridge oscillator.

FIGURE 12-19 Wien bridge circuits.

for instance, we can assume that the voltage gain A is equal to a real constant k:

$$A(j\omega) = k \tag{12-77}$$

Thus the open loop gain $A\beta$ becomes

$$A\beta(j\omega) = \frac{kRC}{3RC + j\omega C^2 R^2 + \frac{1}{j\omega C}} \tag{12-78}$$

Using the Barkhausen criterion and setting the imaginary part of Eq. (12-78) equal to zero, we obtain the frequency of oscillation ω_o:

$$\omega_o = \frac{1}{RC} \tag{12-79}$$

Furthermore, if we set Eq. (12-78) equal to unity, we discover that to have self-sustaining oscillations the values of k should be exactly 3.0.

A simple practical implementation of the Wein Bridge oscillator is shown in Fig. 12-19b. The amplifier used is a two-stage amplifier used in Section 12-2. The feedback resistor R_f is adjusted so that the required gain of 3.0 is obtained. Notice that the high input impedance is obtained by incorporating the biasing resistors as part of the feedback network. Thus, in a practical design, we arrange the parallel combination of R_A and R_B equal to R.

A great many more types of oscillators are possible using various numbers of transistor stages and different types of feedback networks. The analysis techniques for these are generally the same as those given in the previous examples. It is hoped that these examples provide a pattern for the student to follow in analyzing and designing other oscillator circuits. The circuits described were analyzed using low-frequency equivalent circuits. Thus the design equations obtained are generally good for the audio-frequency range. If higher frequency oscillators are desired, it may be necessary to use the hybrid-pi model to analyze these circuits. This is generally more difficult, and, occasionally, more complicated expressions arise for the design equations.

PROBLEMS

12.1 Show that the circuit in Fig. P12-1a can be represented by the block diagram in Fig. P12-1b.

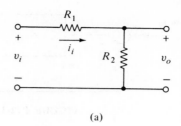

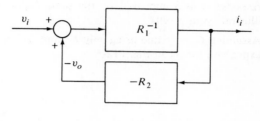

FIGURE P12-1

12.2 Show that the circuit in Fig. P12-2a can be represented by the block diagram in Fig. P12-2b.

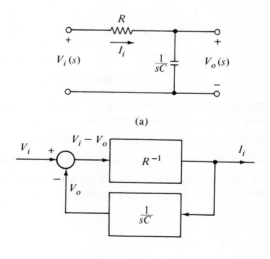

FIGURE P12-2

12.3 The block diagram in Fig. P12-3 represents a unity feedback system (the case when $\beta = 1$). Find the closed loop gain of this system.

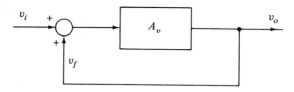

FIGURE P12-3

12.4 In Eq. (12-43) we have shown that high-frequency response is improved by negative feedback. Show by a similar method how the low-frequency response is also improved by the same factor. Use Bode plots to help illustrate your results.

12.5 Assuming C_i large and using only h_{ie} and h_{fe} in the model for T_1, find an expression for R_{in} in Fig. P12-5.

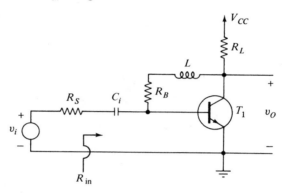

FIGURE P12-5

12.6 Design a phase-shift oscillator to operate at 1 KHz. Specify the minimum value of h_{fe} needed in the transistor.

12.7 Design a Colpitts oscillator to operate at 20 KHz.

12.8 Repeat Problem 12.7 using a Wein Bridge oscillator.

12.9 Obtain the design equations for the Hartly oscillator shown in Fig. P12-9. Neither C_1 or C_2 are large capacitors.

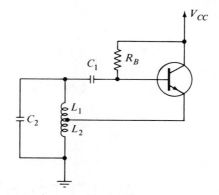

FIGURE P12-9

12.10 Obtain the values for L_1, L_2, C_1, and C_2 in Problem 12.9 for the oscillator to operate at 20 KHz.

REFERENCES

[1] SCHILLING, D. L., and BELOVE, C., *Electronic Circuits: Discrete and Integrated* (McGraw-Hill Book Co., New York, 1968).

[2] CORNING, J. J., *Transistor Circuit Analysis and Design* (Prentice-Hall, Inc., Englewood Cliffs, N.J., 1965).

13

AUDIO POWER AMPLIFIERS

In all the discussions in the previous chapters we have employed small-signal models. This is a valid approach when the signals in a circuit are not too large, that is, when the energy of the signal in an amplifier is a small portion of the energy supplied by the battery or power supply.

The study of power amplifiers usually involves the study of circuits with large signals that are used to operate some type of a mechanical transducer, such as a loudspeaker, pen recorder, or even a motor. For these types of amplifiers, power efficiency becomes quite important. The important question is: How much of the energy supplied by the source is used to operate the mechanical transducers and how much of the energy is dissipated as heat? As a result, graphical techniques as well as experimental techniques are used in the design of power amplifiers. Because of the large signals involved, small-signal models are not generally used.

Figure 13-1 is a block diagram of a typical electronic system in which a power amplifier may be used. It consists of an input transducer, preamplifier, power amplifier, and output transducer. The input transducer generally converts mechanical energy to electrical energy and may be a device such as a phonograph cartridge, microphone, pressure gauge, etc. The amplifier section is split into parts consisting of the preamplifier and the power amplifier. The preamplifier is that portion of the amplifier in which small-signal analysis is used for design. The power amplifier then operates an output transducer such as a loudspeaker, pen recorder, etc., as previously noted. Very often one or more feedback blocks are used with the system shown in Fig. 13-1 and can be connected between any two of these operational blocks.

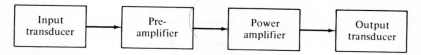

FIGURE 13-1 Block diagram of an electronic system employing a power amplifier.

13.1 Power Considerations

Power calculations are very important in audio amplifiers, since the currents and voltages used will be quite large. If the power dissipation of the transistor is exceeded, then the transistor may burn out. In the earlier chapters it was pointed out that the Q-point in a small-signal amplifier should be below the maximum power dissipation so that the transistor will not be destroyed. This criterion may not be good enough for all types of amplifiers.

The average power dissipated by any device is given by

$$P_{AVG} = \frac{1}{T} \int_0^T v_{AB}(t) i_A(t) \, dt \tag{13-1}$$

where $v_{AB}(t)$ is the total voltage across the device, $i_A(t)$ the total current through it, and T the period of the time-varying portion of v_{AB} or i_A.

Let us now consider the amplifier circuit shown in Fig. 13-2a. We shall assume that the resistor R_E is much smaller than R_L so that very little energy is dissipated in R_E as compared to R_L. The signal power in R_L, or the ac power in R_L, $P_{L,ac}$, can be calculated to be

$$P_{L,ac} = \frac{1}{T} \int_0^T i_c^2 R_L \, dt \tag{13-2}$$

If we further consider the load lines for the amplifier shown in Fig. 13-2b and assume the ac and dc load line to be approximately the same, we can calculate maximum ac power to the load ($P_{L,ac,max}$). The maximum collector ac current becomes

$$i_c \simeq \frac{V_{CC}}{2R_L} \cos \omega t \tag{13-3}$$

and occurs when the Q-point is in the middle of the load lines. The maximum ac power to the load now becomes

$$P_{L,ac,max} = \left(\frac{V_{CC}}{\sqrt{2}\, 2R_L}\right)^2 R_L \tag{13-4}$$

The average power delivered by the power supply P_{CC}, neglecting the

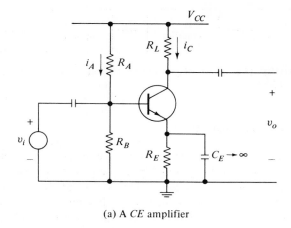

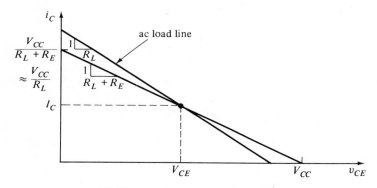

(a) A CE amplifier

(b) The ac and dc load lines for a CE amplifier

FIGURE 13-2 CE amplifier and load line diagram.

biasing current i_A, can be calculated as

$$P_{cc} = \frac{1}{T}\int_0^T V_{cc}i_C(t)\,dt$$
$$= \frac{1}{T}\int_0^T V_{cc}[I_C + i_c(t)]\,dt \qquad (13\text{-}5)$$
$$= V_{cc}I_C$$

The average power dissipated by the transistor P_C can be calculated by subtracting the ac and dc power in the load from the power delivered by the supply:

$$P_C = P_{CC} - \frac{1}{T}\int_0^T i_c^2 R_L \, dt$$

$$= P_{CC} - \frac{1}{T}\int_0^T (I_c + i_c)^2 R_L \, dt \qquad (13\text{-}6)$$

$$= P_{CC} - I_c^2 R_L - \frac{1}{T}\int_0^T i_c^2 R_L \, dt$$

It is interesting to note from Eq. (13-6) that maximum power dissipated in the transistor P_C occurs when the signal current i_c is zero, that is, when no signal is present. This means that maximum collector power dissipation $P_{C,max}$ simply becomes

$$P_{C,max} = V_{CE}I_C \qquad (13\text{-}7)$$

This further implies that, if $P_{C,max}$ is less than P_d, the maximum power dissipation of the transistor (discussed in Chapter 2), the transistor will not be destroyed. This also means that a presence of a signal in this type of amplifier will tend to reduce the power which must be dissipated by the transistor.

The efficiency of an amplifier can be expressed as the ac power in the load divided by power delivered by the supply:

$$\eta = \frac{P_{L,ac}}{P_{CC}} \qquad (13\text{-}8)$$

Since P_{CC} is a constant for this type of amplifier and $P_{L,ac}$, as given by Eq. (13-2), increases with the amplitude, then maximum efficiency η_{max}, when $P_{L,ac}$ is at maximum as given by Eq. (13-4), is

$$\eta_{max} = \frac{P_{L,ac,max}}{P_{CC}}$$

$$= \frac{V_{CC}^2}{8R_L} \Big/ V_{CC}I_C \qquad (13\text{-}9)$$

$$= \frac{V_{CC}^2}{8R_L} \Big/ \frac{V_{CC}^2}{2R_L} = 0.25 = 25 \text{ per cent}$$

Equation (13-9) is obtained by remembering that the maximum output voltage occurs when the Q-point is in the middle of the load line. This allows us to calculate P_{CC} as $V_{CC}^2/2R_L$ by the use of simple geometry. It is interesting to note that the maximum efficiency is only 25 per cent in this type of amplifier.

In most circuits studied thus far, the power levels have been quite low, so that efficiency was not important. However, if several watts or more ac power were required, this amplifier would not be a good choice because of its relatively low efficiency. A more efficient amplifier is the transformer coupled amplifier of Fig. 13-3a.

To calculate the maximum efficiency η_{max} of the transformer coupled amplifier, we shall assume the dc load line to be vertical and the Q-point to be arranged for a maximum voltage swing in v_{CE}, as shown in Fig. 13-3b. We shall assume the losses due to R_A, R_B, and R_E to be negligible as in the previous amplifier and assume the efficiency of the transformer to be 100 per cent.

The power delivered by the supply P_{CC} is constant and is given by

$$P_{CC} \simeq \frac{V_{CC}^2}{R_L'} \tag{13-10}$$

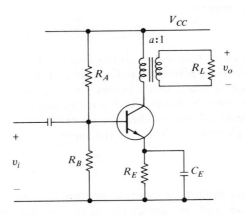

(a) A CE transformer coupled amplifier.

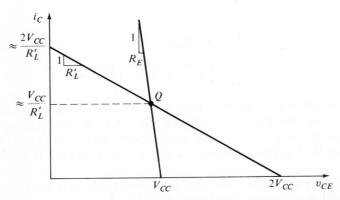

(b) The ac and dc load lines for maximum output voltage swing.

FIGURE 13-3 A CE-transformer coupled amplifier and load line.

The maximum output ac power $P_{L,ac,max}$ can be calculated by assuming the peak-to-peak voltage swing in v_{CE} to be $2V_{CC}$. This output power becomes

$$P_{L,ac,max} = \left(\frac{V_{CC}}{\sqrt{2}}\right)^2 \bigg/ R'_L$$
$$= \frac{V_{CC}^2}{2R'_L} \quad (13\text{-}11)$$

where
$$R'_L = a^2 R_L$$

Thus the maximum efficiency of this amplifier becomes the ratio of Eqs. (13-11) and (13-10):

$$\eta_{max} = \frac{P_{L,ac,max}}{P_{CC}} = \frac{1}{2} = 50 \text{ per cent}$$

Thus we see that maximum efficiency of a transformer coupled amplifier is 50 per cent. This is an improvement over the ordinary H-type biased amplifier but is not yet the optimum that can be obtained.

It is also easy to see, as in the previous case, that the maximum collector dissipation of the transformer coupled amplifier occurs when there is no signal present, as given by

$$P_C = V_{CE} I_C = \frac{V_{CC}^2}{R'_L} \quad (13\text{-}12)$$

Also, as before, this means that Q-point should be in the maximum power dissipation hyperbola, as specified for the transistor to be used in the circuit.

13.2 Classification of Power Amplifiers

Power amplifiers are classified according to that portion of the input sine wave signal cycle during which output or load current flows. For example, in all the amplifiers studied thus far, output current (usually collector current) flows during the entire input sine wave cycle. This type of amplifier is said to be a *Class A* amplifier, or it is an amplifier operating in the Class A mode. See Fig. 13-4a.

A *Class AB* amplifier in which output current flows for more than one half of the cycle, but less than a full cycle, as shown in Fig. 13-4b. Figure 13-4c illustrates the action of a *Class B* amplifier, in which the output current flows exactly for one half of a cycle. Finally, Fig. 13-4d illustrates the operation of a *Class C* amplifier. In a Class C amplifier the output current flows for less than one half of the input cycle.

Class A amplifiers are generally used in small-signal amplifiers but can be used in low-powered power amplifiers. The disadvantage of this class of amplifier is the relatively low efficiency. Class B and AB amplifiers are more common for audio power amplifiers because of their higher efficiency. Class C amplifiers are used primarily in radio-frequency amplifiers.

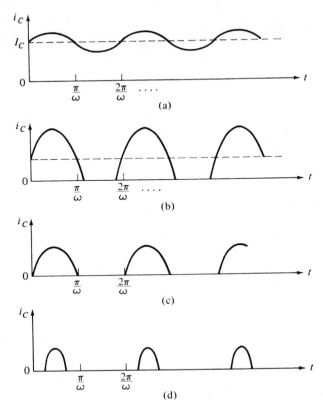

FIGURE 13-4 Graphs indicating flow of i_C as function of time for various classes of amplifiers: (a) Class A, (b) Class AB, (c) Class B, (d) Class C.

13.3 Class B Push–Pull Amplifiers

In Section 13-1 we discovered that the maximum efficiency of a Class A transformer coupled amplifier was only 50 per cent. This was partly due to the fact that the collector current flowed at all times, even if no signal was present. The use of a Class B amplifier increases the efficiency by having the collector current flow only when a signal is present.

Fig. 13-5 is an example of a very elementary *push–pull amplifier* operating in a Class B mode. The name push–pull in fact describes the action of the circuit. The operation of this circuit is quite simple. When v_i is zero, no current i_{B1} or i_{B2} flows, and hence the output current i_L is also zero. When v_i becomes positive, a current i_{B1} is able to flow, since T_1 acts as an emitter follower resulting in current i_{C1}, h_{FE1} times larger than i_{B1}. The voltage across

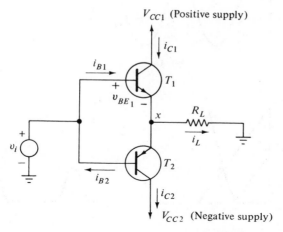

FIGURE 13-5 Elementary push–pull amplifier.

R_L is approximately equal to the input. During the positive portion of v_i the transistor T_2 is cut off and no current flows. During the negative portion of v_i, T_2 acts as an emitter follower and T_1 is cut off. Figures 13-6a through 13-6e help to illustrate the push–pull action of this amplifier graphically. To obtain the effect of v_i on v_{BE1}, we must use load-line analysis by recognizing that the input voltage v_i in Fig. 13-5 is given by

$$v_i = v_{BE1} + i_{B1} h_{FE1} R_L \tag{13-13}$$

This can be written in the form suitable for load-line analysis as

$$i_{B1} = -\frac{1}{h_{FE1} R_L} v_{BE1} + \frac{1}{h_{FE1} R_L} v_i \tag{13-14}$$

Equation (13-14) is the equation of a straight line with a negative slope of $1/h_{FE1} R_L$ and a v_{BE1} intercept equal to the instantaneous value of v_i. This means that the load line moves back and forth across the v_{BE1}, i_{B1} graph with v_i. The intersection of the load line with this curve gives the resuling $i_{B1}(t)$, as shown in Fig. 13-6b. Notice the nonsinusoidal appearance of i_{B1} due to the nonlinearity of the input characteristics.

Figure 13-6c shows the collector current i_{C1} as being approximately the same shape as i_{B1}, since i_{C1} is given by

$$i_{C1} = h_{FE1} i_{B1} \tag{13-15}$$

if h_{FE1} is approximately constant.

In Fig. 13-6(d), i_{C2} is shown, and here we see that i_{C2} flows during the negative half cycles of v_i. Since i_L is given by

$$i_L = i_{C1} - i_{C2} \tag{13-16}$$

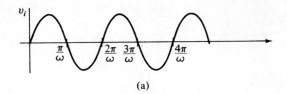
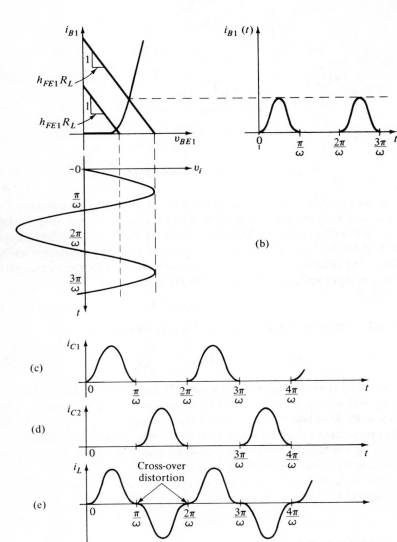

FIGURE 13-6 Voltages and currents in an elementary push–pull amplifier (Fig. 13-5). (a) The input voltage v_i. (b) The effect of v_i on i_{B1}. (c) The collector current of T_1. (d) The collector current of T_2. (e) The load current $i_L = i_{C1} - i_{C2}$.

by applying Kirchhoff's current law at node x in Fig. 13-5, we can sketch the load current i_L as shown in Fig. 13-6e.

Because of the nonlinearity of the input characteristics, there is distortion on the zero crossings of i_L at $t = 0, \pi/\omega, 2\pi/\omega, \ldots$. This type of distortion is called *crossover distortion* and can be practically eliminated by forward-biasing T_1 and T_2 slightly into Class AB operation. Figure 13-7 illustrates a push–pull amplifier operating in Class AB.

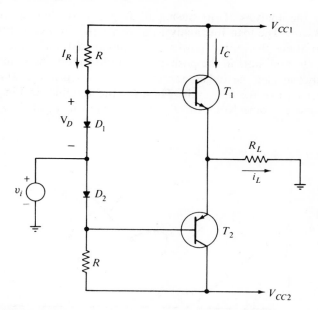

FIGURE 13-7 An elementary push–pull amplifier forward biased to eliminate crossover distortion.

The forward bias of the transistors T_1 and T_2 in Fig. 13-7 is achieved by the use of the two diodes D_1 and D_2. Resistors can be used instead of the diodes, but the diodes provide temperature stability. It can be shown experimentally that for an *npn* transistor V_{BE} decreases with increasing temperature according to the formula

$$\frac{\Delta V_{BE}}{\Delta T} \simeq -2.5 \text{ mV/}°\text{C} \tag{13-17}$$

The voltage across a diode V_D varies at about the same rate with temperature. By keeping D_1 and D_2 at the same temperature as T_1 and T_2 in Fig. 13-7 (this can be done by physical contact), a constant current I_C through T_1 and T_2 is maintained with temperature.

I_C can be increased by decreasing R, hence the current through the diode. This usually decreases the amount of crossover distortion, as the amplifier operates more into Class AB. The correct amount of I_R, and thus I_C, is usually found experimentally by measuring the crossover distortion. A compromise between the minimum value of I_C and the mimimum distortion is usually made.

13.4 Load Line Analysis of Class B Amplifiers

To simplify the analysis of the push–pull amplifier we shall consider only the Class B case. The load line analysis, as in the previous case, can lead to information about the output power, power dissipated, and efficiency.

Since both transistors in the push–pull amplifier act in the same manner, we need only consider one and assume that the other acts in the same fashion for the other part of the half cycle. Thus, for the amplifier in Fig. 13-5, we shall only concern ourselves with T_1. A load line for T_1 is shown in Fig. 13-8.

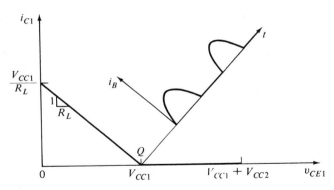

FIGURE 13-8 Load line for T_1 (Fig. 13-5).

Notice that the load line is made up of two segments. The first segment is for the positive half cycles when T_1 is conducting and consists of load line of slope $(-1/R_L)$ and a v_{CE1} intercept of V_{CC1}. The second segment is for the negative half cycles when i_C and i_B are zero and lie on the $i_{C1} = 0$ axis between V_{CC1} and $V_{CC1} + V_{CC2}$. The Q-point is situated at the intersection of these two curves. For a Class AB amplifier it would be situated higher up the positive half-cycle portion.

The power delivered by the supply V_{CC1} is

$$P_{CC1} = V_{CC} \frac{1}{T} \int_0^T i_{C1}(t)\, dt \qquad (13\text{-}18)$$

If $i_{C1}(t)$ during the positive half cycle is given by

$$i_{C1}(t) = I_{C,max} \sin \omega t \qquad (13\text{-}19)$$

Eq. (13-18) becomes simply

$$P_{CC1} = V_{CC1}\frac{I_{C,max}}{\pi} \qquad (13\text{-}20)$$

If the two power supplies in the push–pull amplifier are the same voltage, the total power supplied by the two power supplies P_{CC} becomes

$$P_{CC} = \frac{2}{\pi}V_{CC}I_{C,max} \qquad (13\text{-}21)$$

where $\qquad V_{CC1} = -V_{CC2} = V_{CC}$

The maximum power supplied by the supplies $P_{CC,max}$ becomes

$$\begin{aligned} P_{CC,max} &= \frac{2}{\pi}V_{CC}\frac{V_{CC}}{R_L} \\ &= \frac{2}{\pi}\frac{V_{CC}^2}{R_L} \end{aligned} \qquad (13\text{-}22)$$

The power delivered to R_L, P_L, is

$$P_L = \tfrac{1}{2}I_{C,max}^2 R_L \qquad (13\text{-}23)$$

The maximum power to the load $P_{L,max}$ becomes

$$P_{L,max} = \frac{V_{CC}^2}{2R_L} \qquad (13\text{-}24)$$

The power dissipated in the transistors T_1 and T_2, $2P_C$, is the difference between the power delivered by the supply and the power delivered by the load:

$$2P_C = P_{CC} - P_L \qquad (13\text{-}25)$$

If Eqs. (13-21) and (13-23) are substituted into Eq. (13-25), we obtain

$$P_C = \frac{2}{\pi}V_{CC}I_{C,max} - \frac{1}{2}R_L I_{C,max} \qquad (13\text{-}26)$$

If we differentiate Eq. (13-26) with respect to $I_{C,max}$ to obtain the collector current for which P_C is at a maximum, we obtain

$$I_{C,max} = \frac{2}{\pi}\frac{V_{CC}}{R_L} \qquad (13\text{-}27)$$

If Eq. (13-27) is substituted into Eq. (13-26), we obtain the maximum collector dissipation for both transistors $2P_{C,max}$ to be

$$2P_{C,max} = \frac{2}{\pi^2}\frac{V_{CC}^2}{R_L} \qquad (13\text{-}28)$$

This means that each transistor must dissipate at least the following energy:

$$P_{C,max} = \frac{1}{\pi^2}\frac{V_{CC}^2}{R_L} \qquad (13\text{-}29)$$

If we now compare the maximum collector dissipation as given by Eq. (13-29) with the maximum output power in Eq. (13-24), we see that

$$P_{C,max} \simeq \tfrac{1}{5} P_{L,max} \qquad (13\text{-}30)$$

This means that, if we want to build a Class B push–pull amplifier that is capable of delivering say 5 W, each transistor must dissipate only 1 W each. Equation (13-30) is a good rule-of-thumb formula to remember when designing power amplifiers.

The efficiency η can be calculated from Eqs. (13-21) and (13-23):

$$\eta = \frac{P_L}{P_{CC}} = \frac{\tfrac{1}{2} I_{C,max}^2 R_L}{\dfrac{2}{\pi} V_{CC} I_{C,max}}$$

$$= \frac{\pi}{4} \frac{I_{C,max} R_L}{V_{CC}} \qquad (13\text{-}31)$$

The maximum efficiency η_{max} is obtained when $I_{C,max}$ is at a maximum:

$$\eta_{max} = \frac{\pi}{4} \simeq 78.5 \text{ per cent} \qquad (13\text{-}32)$$

It is interesting to note that this efficiency is considerably higher than the previous amplifiers studied.

13.5 Practical Push–Pull Amplifiers

The amplifier shown in Fig. 13-9 is a practical implementation of a push–pull amplifier. The circuit consists of a pair of compound emitter followers $T_3 - T_5$ and $T_2 - T_4$, and a single-stage CE amplifier T_1.

The compound emitter followers are necessary at large-signal levels, since the current gain of a single transistor would not be high enough and would load the voltage amplifier, which precedes the Class B push–pull emitter followers. The combination of T_3-T_5 and T_2-T_4 can be considered to be Darlington connections, since at high current levels the input impedances of T_4 and T_5 are lower than the values of R_6 and R_7, so that R_6 and R_7 can be thought to be an open circuit. The resistors R_6 and R_7 are necessary to forward-bias T_4 and T_5 into Class AB by equal amounts. This is accomplished, since approximately the same Q-point current flows through R_6 and R_7. The compound combination is slightly forward-biased by D_1, D_2, D_3, and R_4. It may be necessary in a practical case to select R_4 in order to eliminate crossover distortion.

Biasing for the voltage amplifier T_1 is accomplished by R_1 and R_2, which also furnishes ac feedback from the output. The correct value of R_1 is determined experimentally. The setting of R_1 determines the amount of collector current of T_1 and hence the voltage drops across the diodes. Increasing the

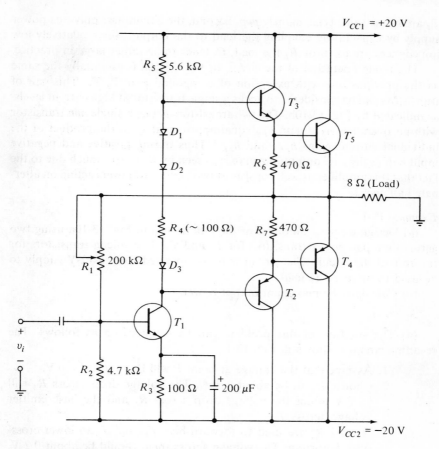

FIGURE 13-9 A practical implementation of a push–pull amplifier.

collector current in T_1 (accomplished by decreasing R_1) increases the current in the output transistors.

This amplifier can be generally made to work satisfactorily by choosing supply voltages (V_{CC1} and $-V_{CC2}$) anywhere between about 6.0 and 30 V. The theoretical maximum input power can be determined using Eq. (13-24). In this case, the maximum output power turns out to be 25 W. This means that T_4 and T_5 should be capable of dissipating 5.0 W as given by Eq. (13-30). Transistors T_1, T_2, and T_3 can be conventional small-signal transistors, and T_4 and T_5 should be power transistors, mounted on a good heat sink to keep the transistors at a safe operating temperature.

Another practical implementation of a Class AB push–pull amplifier is shown in Fig. 13-10a. There are several essential differences in this circuit when compared to the one of Fig. 13-9. First, both power transistors T_4 and

T_5 are of the same type, namely *pnp*. Second, the circuit uses only one power supply by capacitively coupling the load to the output. Since relatively low impedances are used for R_L the load, C, tends to be rather large in practice.

The mode operation of the circuit in Fig. 13-10a is essentially the same as the previous one with exception of compound pair T_3-T_5. This pair of transistors can be considered to act as a single transistor at high current levels, as indicated by Fig. 13-10b. The two transistors act as a single *npn* transistor with an overall current gain h_{FE} equal approximately to the product of the individual current gains h_{FE3} and h_{FE5}. Thus during positive and negative input half cycles, the output impedance as seen by R_L is very much due to the fact that this amplifier as well consists of two emitter followers acting on alternate half cycles.

Example 13-1:

(a) Design a power amplifier of the type shown in Fig. 13-10a using two germanium *pnp* power transistors for T_4 and T_5. Use silicon transistors for the rest of the circuit. The circuit is to be operated from a 12-V supply to be used to drive a 4-Ω load.*

(b) Calculate the maximum output power.

Solution:

(a) The solution of this problem can be summarized as follows. The resulting circuit is shown in Fig. 13-11.

1. Assume that the voltage at point P will be $V_{CC}/2 = -6$ V.
2. Choose I_{C1} to be about 1 mA. The voltage drop across R_3 will be 6 V minus the voltage drop across R_1 and the base emitter voltage across T_2.
3. R_1 and R_2 are used to forward-bias T_4, and T_5 to lower crossover distortion. The voltage across them should be about 0.2 V.
4. Hence

$$R_3 = \frac{6 - 0.8}{1 \times 10^{-3}} = 5.2 \text{ k}\Omega$$

5. The current through R_4 is I_{C1}/h_{FE1}, and voltage drop across it is 6–0.6 V. For an h_{FE1} of 100 for T_1 we have

$$I_{B1} = \frac{1 \text{ mA}}{100 \, \Omega} = 10 \, \mu\text{A}$$

$$R_4 = \frac{5.4 \text{ V}}{10 \, \mu\text{A}} = 540 \text{ k}\Omega$$

6. For $I_{C2} = 1$ mA,

$$R_1 = R_2 = 200 \, \Omega$$

*An amplifier of this type could be used with a preamplifier such as that of Fig. 12-7.

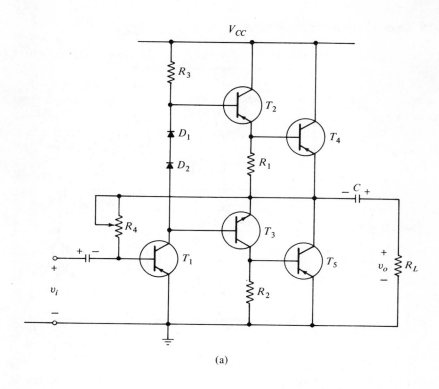

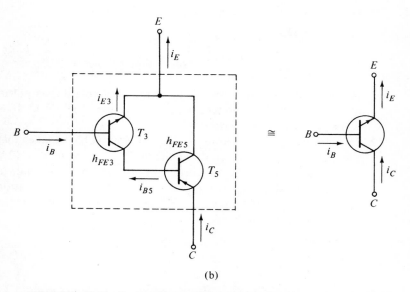

FIGURE 13-10 Push–pull amplifier: (a) Basic class AB push–pull amplifier (b) Darlington equivalent of T_3 and T_5.

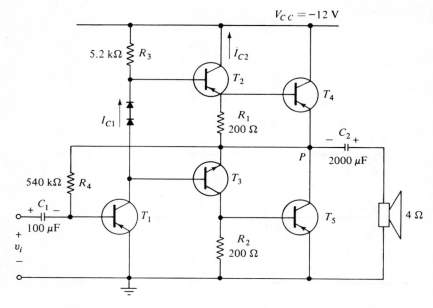

FIGURE 13-11 Push–pull amplifier in example problem 13-1.

7. C_2 should be large enough to provide coupling at the lowest frequency to be used. Thus, to obtain a 3-dB-down point at 20 Hz,

$$4\,\Omega = \frac{1}{2\pi \cdot 20 C_2}$$

$$C_2 = \frac{1}{2\pi(20)(4)} = 2{,}000\ \mu\text{F at 12 WVDC}$$

8. Similarly, C_1 should be sufficiently large to couple the output of a preamplifier to a common emitter amplifier. 100 μF at 10 V would probably be sufficient.
9. In choosing T_4, T_5, the power transistors, care should be taken that the maximum collector emitter voltage BV_{CEO} be at least -12 V and the maximum collector current be at least 1.5 A, as obtained by load line analysis (see Fig. 13-12). The power dissipation of T_4 and T_5 should be at least one fifth the power output obtained in part (b) of this problem.

(b) The maximum output power $P_{L,max}$ becomes

$$P_{L,max} = \frac{\left(\frac{V_{CC}}{2\sqrt{2}}\right)^2}{R_L} = \frac{V_{CC}^2}{8 R_L}$$

$$= \frac{12^2}{32} \simeq 4.5\ \text{W}$$

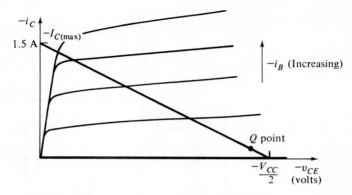

FIGURE 13-12 Load line analysis for T_4 and T_5 of Ex. 13-1.

PROBLEMS

13.1 (a) In the circuit shown in Fig. P13-1 calculate the dc power dissipated in R_L.
    (b) With no signal present, what will be the power dissipated by the transistor?
    (c) What is the maximum ac power that is dissipated by R_L? The ac signal must remain undistorted.
    (d) What will be the collector dissipation with the circuit operating with maximum ac power into R_L?

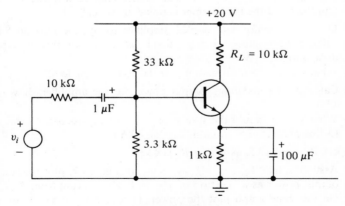

FIGURE P13-1

13.2 The circuit shown in Fig. P13-2 is called an inductively coupled amplifier.
    (a) Calculate the value of L necessary such that the lower frequency cutoff of the amplifier is at least ω_L. Assume C_o, C_i, and C_E to be very large capacitors.
    (b) Bias an amplifier such as this with $V_{CC} = 12$ V and $I_C = 10$ mA. Draw the ac and dc load lines for this amplifier. Assume R_L to be 1.0 kΩ.

358 Chap. 13 / Audio Power Amplifiers

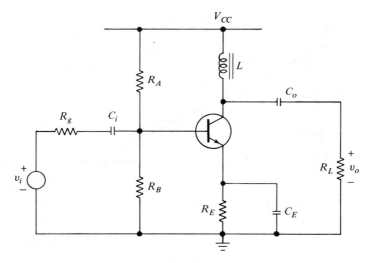

FIGURE P13-2

(c) What will be the maximum peak-to-peak output voltage v_o across R_L for the circuit as biased in part (b)?

13.3 (a) Calculate the maximum theoretical efficiency for an amplifier of the type shown in Fig. P13-2. Neglect the power loss due to R_E and the biasing resistors R_A and R_B.

(b) How does the efficiency of this amplifier compare with the *RC* coupled amplifier and the transformer coupled amplifier?

13.4 (a) Design a transformer coupled amplifier to operate into an 8-Ω load using a 12-V supply and a silicon *pnp* transistor that is capable of dissipating 0.25 W.

(b) What will be the maximum power delivered to the load?

13.5 (a) Calculate the maximum output power for the amplifier shown in Fig. 13-9.

(b) What power should the transistors T_4 and T_5 dissipate? (The answers to this problem are given in Section 13.5.)

13.6 Repeat Problem 13.5, assuming R_L is 4 Ω.

13.7 (a) Assuming that $V_{CC1} = -V_{CC2} = V_{CC}$ in Fig. 13-7, plot a graph of the output power as a function of V_{CC} with V_{CC} varying from 5 to 50 V.

(b) On the same graph plot the power dissipated by either T_1 or T_2 as a function of V_{CC}.

13.8 (a) Design an amplifier of the type shown in Fig. 13-10a to operate from a negative 24-V supply into an 8-Ω load. Use silicon transistors throughout the design.

(b) What will be the output power of this amplifier and what should the power dissipation of the two output transistors be?

13.9 Design a suitable power supply to operate the amplifier in Problem 13.8.

INDEX

A

ABCD parameters, *see* cascade parameters
ac, 24, 26
  equivalent circuits, 39-40, 44-47, 100-107
air-core transformer, 225
alpha cutoff frequency, 249
amplifiers, 21, 25
  audio, 315-320, 340-341, 352-357
  bootstrapped, 136-142
  cascaded, 142-151
  cascode, 143-147, 270-273
  class A, 345-346
  class AB, 345-346, 349
    push-pull, 349, 352
  class B, 345-346
    push-pull, 345-350, 357
  class C, 345-346
  common base (CB), 115-123, 143, 264-269
  common collector (CC), 128-135, 143, 273-278, 311-313
  common drain, 171-175

Amplifiers (*Cont*)
  common emitter (CE), 53-81, 109-112, 143, 203-217, 260-264, 342-343
  common gate, 175-176
  common source, 168-170, 217-221, 278-280
  current feedback, 321-326
  emitter follower, *see* common collector
  feedback, 315-317, 323-326
  power, 345-357
  push-pull, 346-347
  source follower, *see* common drain
  specifications, 182
  transformer coupled, 229-233, 344-345
  two-stage, 122-128
  unbypassed emitter resistor, 76-80, 143
  voltage feedback, 315-320
approximate sources:
  current, 8
  voltage, 8

B

Barkhausen stability criterion, 327-328
base, 16, 97-99
base bias, 52-53 (*see also* constant current biasing)
basic feedback system, 310
beta β, 103-108, 183, 185-186
beta cutoff frequency, 247
bias, 27
  base, 16, 97-99
  reverse, 88, 98
biasing, 27, 52-59
  of the CB amplifier, 115-116
  of the CC amplifier, 128
  of the CE amplifier, 27, 29-31, 52-59
  constant current, 52-54
  current feedback, 58-59
  of the FET, 164-168
  H-type, 55-58
bipolar junction transistor, BJT, 155
block, 309
Bode frequency response plots:
  of beta, 186, 248
  of the CB amplifier, 211, 269-270
  of the CC amplifier, 211, 273-278
  of the CE amplifier, 203-217, 256-264
  of FET amplifiers, 217-222, 278-280
  of general networks, 192-200
  of the high-pass R-C filter, 201-202
  transformer coupled amplifier, 231-232
  transformers, 226-229
Boltzmann's constant, 90
bootstrapped amplifiers, *see* amplifiers
breakdown region:
  of transistors, 183-184
  of zener diodes, 296
breakdown voltage:
  of transistors, 183-184
  of zener diodes, *see* zener voltage
break frequencies, 192-194
break point, *see* break frequencies
bypass capacitor, 56, 205-209, 214-222

C

capacitive coupling, 209-211
cascade parameters, 3
cascading, 122-128, 130-131, 142-151
cascode amplifier, 144-147, 270-273
  high-frequency response, 270-273
chain-type biasing, 55-57 (*see also* H-type biasing)
characteristics:
  diode, 89-93, 101
  transistor:
    dynamic input, 43-47
    dynamic output, 45-47, 68-73
    field effect (FET), 158-163
    static input, 17, 19-20, 39, 42
    static output, 45-47, 68-73
  zener diode, 296-297
circuit specification, 181-182
classification of power amplifiers, 345-346
closed-loop gain, 310-311, 314-317, 321-326
coefficient of coupling:
  transformer, k, 226
collector, 16, 97-99
collector breakdown voltage, *see* maximum collector breakdown voltage BV_{ceo}
common base amplifier, *see* amplifiers, common base
common collector amplifier, *see* amplifiers, common collector
common drain amplifier, *see* amplifiers, common drain
common emitter amplifier, *see* amplifiers, common emitter
common gate amplifier, *see* amplifiers, common gate
common source amplifier, *see* amplifiers, common source
conduction band, 86
constant current biasing, 52-53, 97, 116-117
conventional current, 21

corner frequency, *see* break frequency
coupling (*see also* cascading):
  direct, 130, 142-151
  R-C, 142-144, 214-222, 341-343
  transformer, 222-233, 344-345
crossover distortion, 349
current amplification factor, forward, 17, 28
  h_{FE}, 17, 53
  h_{fe}, 28, 40-48, 53
    dependence on temperature, 53
current division, 7
current feedback, *see* biasing, current feedback
current feedback amplifier, *see* amplifiers, current feedback
current source, *see* sources, ideal current

D

Darlington connection, 149, 355
dc, 24
decibel, 188-189
  db, 188
  dBv, 188-189
degenerative feedback, 326
depletion mode, 160
depletion region, 157
differential gain, 148
diffusion, 88
diffusion capacitance, 246
direct coupling, 123, 126-128, 130, 144-151
doping, 88
drift, 88

E

efficiency:
  of class A amplifiers, 341-343
  of power amplifiers, 350-352
  of transformer coupled amplifiers, 344-345

electronic system, 309
emitter, 16, 97-99
emitter biasing, 55 (*see also* biasing, *H*-type)
emitter bypass capacitor, 56, 205-209
emitter follower, *see* amplifiers, common collector
emitter follower voltage regulator, 293-296, 302-306
  load line analysis of, 294
enhancement mode, 160
equivalent circuits:
  BJT transistor in biased state, 60-63, 75, 76, 109-111, 118-121, 131-132
  equivalent tee:
    common base, 101, 106
    common emitter, 102-108
  field effect transistor (FET), 161-164
  generic, 99-103
  hybrid (h), 35-47, 104-108
  hybrid-pi, 239-246
  r-parameter, 104-108
  total, 99-100
  z-parameter, 104-108
extrinsic conduction, 88

F

f_T, *see* gain bandwidth product
feedback gain, 310
field effect transistor, 155-159 (*see also* amplifiers)
  equivalent circuit, *see* equivalent circuits
  high-frequency response, 278-280
  low-frequency response, 217-222
forward gain, 310, 312
frequency response plots, *see* Bode frequency response plots
fullwave bridge rectifier, 288
fullwave rectifier, 287

G

gain bandwidth product, 185, 246-249, 258
generic model, *see* equivalent circuits, generic
germanium transistors, 29, 188
g-parameters, *see* inverse hybrid parameters

H

h_{fe}, 17, 21, 23, 28, 56
h_{ie}, symbol for short-circuit input impedance, 28, 38, 40-48, 106, 107
h_{oe}, symbol for open-circuit output impedance, 38, 40-48, 104, 107
h_{re}, symbol for open-circuit reverse voltage transfer ratio, 38, 40-48, 104-105
half-wave rectifier, 285-286
high-frequency cutoff, 205
hole, 86
h-parameters, 187 (*see also* equivalent circuits, *h*-parameter)
H-type biasing, *see* biasing, *H*-type
hybrid equivalent circuit, 35-40
hybrid parameters, 35, 38, 104-108
hybrid-pi equivalent circuit, *see* equivalent circuits, hybrid-pi

I

ideal current source, *see* sources, ideal current
ideal diode, 92-93
ideal voltage source, *see* sources, ideal voltage
IGFET, 155, 159-161
impedance functions, 189
incremental drain resistance, 161-163
input, 12
 characteristics, 27, 39, 161-162
 impedance, 12-14

Input (*Cont*)
 terminal, 12, 48
input coupling capacitor, 25-29, 203-205
 for bipolar junction transistor amplifiers, 25-29, 203-207, 214-217
 for field effect transistor amplifiers, 217
input impedance, 12-14
 of bipolar junction transistor amplifiers, 143
 of a CB amplifier, 116-119
 of a CC amplifier, 131-132
 of a CE amplifier, 28
  using *h* parameters, 63-67
  using *r* parameters, 110-111
 of a FET amplifier, 170-176
 of an ideal transformer with resistive termination, 224
 of an unbypassed emitter resistor amplifier, 77
insulated gate field effect transistor, *see* IGFET
intrinsic conduction, 35, 86
inverse cascade parameters, 35
iron-core transformer, 225, 227

J

junction, 88
junction diode, 88
junction field effect transistor, 155-159 (*see also* field effect transistor)

L

leakage current, 89, 91, 100, 186-187
 I_{CBO}, 98-99, 186
 I_{CEO}, 187
 I_{CO}, 89
loading, 124-128
load line, 68-73
 for design, 137, 185, 357

load line analysis, 68-72, 137
  for CC amplifiers, 135-136
  for CE amplifiers, 68-73
  for emitter follower voltage regulator, 294
  for FET amplifiers, 167
logarithmic gain, *see* decibel
low-frequency cutoff, 204

M

maximum collector breakdown voltage BV_{CEO}, 185
maximum collector current $I_{C\max}$, 185
maximum collector power dissipation, P_d, 18-20, 183
maximum temperature (of transistors), 185
measurements of:
  input impedance, 13
  h_{ie}, 42, 44
  h_{fe}, 41, 45-46
  h_{oe}, 41, 47-48
  h_{re}, 42, 46
  output impedance, 13-14
metal oxide semiconductor field effect transistor (MOSFET), *see* IGFET
mid-band gain, 204, 262
Miller effect, 251-253
  in CB amplifiers, 264
  in CE amplifiers, 253-257
  in FET amplifiers, 278
MOSFET, *see* IGFET

N

n-channel FET, 156
negative feedback, 315, 316-320, 323, 326
network theorems:
  Norton's, 6, 10
  superposition, 10-12
  Thévenin's, 3, 9

noise, 187
non-linear resistor, 43
npn, 20, 97-98
n-type material, 88
Nyquist stability criterion, 326-327

O

open-circuit impedance parameters, 34-35, 46-49
open-circuit voltage, v_{oc}, 6
open-loop gain, 310
output characteristics:
  of bipolar junction transistors, 18, 39
  of field effect transistors, 159-160
output impedance:
  of the CB amplifier:
    using h parameters, 143
    using r parameters, 110-111, 143
  of the CC amplifier:
    using h parameters, 133-134, 143
    using r parameters, 132-133, 143
  of the CE amplifier:
    using h parameters, 120-121, 143
    using r parameters, 110-111, 143
  of the common gate amplifier, 175
  of the common source amplifier, 168-170
  with current feedback, 323
  with voltage feedback, 315

P

p-channel FET, 156
pentode operating region, 176
phasors, 24
pinch off voltage, 157
pnp, 20, 97-98, 245-246
positive feedback, 315, 327
port, 33
power dissipation, 18, (*see also* efficiency)

practical oscillators, 329-336
  Colpitts, 333
  phase shift, 329-333
  Wien bridge, 333-336
practical sources:
  construction of, 9
  current source, 7-10
  Norton equivalent of, 10
  Thévenin equivalent of, 9
  voltage source, 7-9
p-type material, 88
push-pull amplifiers, *see* amplifiers, push-pull

Q

Q-point, 27, 29, 48
  graphical solution, 58, 68, 70
  selection for design, 136, 183-185

R

R-C coupling, 142, 144, 222
R-C filter, 290
  high-pass, 201-202
R-C filtered rectifier, 290-293
regenerative feedback, 326
regulated power supply:
  design procedures, 298-302, 303-306
  emitter follower type, 293-296
  zener diode type, 296-301
regulation def., 289
response, 10
reverse biasing, 88
ripple def., 289

S

saturation current, 157
self bias, 64
semiconductor diode, 85-92
short-circuit admittance parameters, 35
short-circuit current, i_{sc}, 6

signal, 12, 22, 182
silicon transistor, 22, 188
slope, 41-43
small signal models, *see* equivalent circuits
small signal voltage, 24, 39
source follower, *see* amplifiers, common drain
sources, (*see also* practical sources):
  ideal current, 1-3
  ideal voltage, 1-3
static input characteristics, 17, 47
static output characteristics, 17, 46
step-down transformer, 229
summing junction, 309
superposition, *see* network theorems
symbolic notation for transistors, 24
symbols:
  for a bipolar junction transistor, 20
  for a diode, 89
  for a field effect transistor, 156
  for an IGFET, 160
  for a MOSFET, 160
  for a zener diode, 297
system, 309

T

tee model, *see* equivalent circuits, r-parameter
Thévenin equivalent circuit, 3-6, 65
tolerances, 3
transconductance, 161
transfer function, 189
transformers, 46-47, 222-229, 286-288, 305
transforms, 189
transistor specifications, 182-188, 247-251
tremolo, 177
triode operating region, 176
two-ports, 12, 33-40, 189
two-stage amplifiers, *see* amplifiers, two-stage

U

unbypassed emitter resistor amplifier, 75-79, 143

V

v_T, 28
voltage amplifier, 25, 29
voltage division, 7
voltage gain A_v, 29
  of the cascode amplifier, 145, 270-273
  of the CB amplifier:
    at high frequencies, 264-268
    using h parameters, 143
    using r parameters, 109-111, 143
  of the CC amplifier, 128-131, 311-312
    at high frequencies, 273-278
    using h parameters, 131-132, 143
    using r parameters, 143
  of the CE amplifier:
    at high frequencies, 253-258
    using h parameters, 120-122, 143
    using r parameters, 118-120, 143
  of the common gate amplifier, 175
  of the common source amplifier, 168-169
  of the difference amplifier, 147-148
  of FET amplifiers at high frequencies, 278-280
  with current feedback, 324-326
  with voltage feedback, 313-315
voltage feedback system, 312

Y

y-parameters, *see* short-circuit admittance parameters

Z

zener diode regulator, *see* regulators, zener diode
zener voltage, 296
zero bias, 167
z-parameters, *see* open-circuit impedance parameters